How to
Pass the SAS
Selection Course

How to
Pass the SAS
Selection Course

Chris McNab

SIDGWICK & JACKSON

First published 2002 by Sidgwick and Jackson
an imprint of Pan Macmillan Publishers Ltd
20 New Wharf Road, London N1 9RR
Basingstoke and Oxford

Associated companies throughout the world

www.macmillan.com

ISBN: 0 283 07343 8 ⭕

1 3 5 7 9 8 6 4 2

A CIP catalogue record for this book is available from the British Library.

Project Editor: Charles Catton/Conor Kilgallon
Designer: Brian Rust
Picture Research: Lisa Wren

Printed in Italy

Eurolitho S.p.A., Cesano Boscone (MI)

Contents

CHAPTER 1

The
Regiment

Rarely has a military unit so captured the public imagination as the SAS. Twice a year, about 150 hopefuls arrive at Hereford intending to join the ranks of the elite. Yet by the end of Selection, less than 15 soldiers will remain.

Colonel Charles Beckwith, founder of the US Army's elite counter-terrorist unit 1st SFOD-D (Delta Force), once commented about the Special Air Service: 'There is the SAS and there is everyone else'. Of all the world's elite military units, the SAS occupies the summit of the hierarchy. Its reputation for endurance, ferocity, and intelligence in combat is legendary.

For all that we know about the SAS, and for all the public interest in its exceptional personnel, there remains a gap between perception and reality concerning the Regiment. Many soldiers, magnetized by a life in the SAS, are attracted by the possibility of acquiring the status that comes with belonging to a unit which cannot be surpassed. Yet, such people are not the ideal candidates for the SAS. Those who wear the SAS beret are not in the business of flexing their egos, and people who boast about their elite status usually do not have the discipline or discretion that comes with elite unit membership. The SAS is looking for a quite different person.

LEFT: An SAS team prepare to demolish a window prior to entering the Iranian Embassy during Operation Nimrod, 1980. The torches mounted on their MP5 sub-machine guns can be used to blind opponents momentarily and also assist rapid aiming in smoke-filled rooms.

LEFT: Possibly the most famous elite unit badge in the world. The SAS badge was reputedly designed by Corporal Bob Tait and was approved for official use by General Auchinleck in 1942. The sword represents King Arthur's Excalibur while the shades of blue are the colours of Oxford and Cambridge.

such as North Africa and Malaysia, but the general public were generally unaware that such a proficient military force was in existence. All that changed on 5 May 1980.

The remarkably efficient operation at the Iranian Embassy in Princes Gate, London, took the SAS from obscure unit to household name almost in the same time as the 17 minutes it took them to annihilate their opponents and secure the building. Though the operation has been well documented, it is worth looking once more at what went on, and evaluating how it affected the perception of what it is actually like to be in the SAS.

The Princes Gate siege began on the morning of 30 April 1980. The protagonists were, apart from the SAS, five terrorists ostensibly of the Democratic Front for the Liberation of Arabistan (DFLA), although evidence suggests that they could have been Iraqi agents. The DFLA were a radical organization seeking autonomy for an Arabic-speaking region of Iran, Arabistan, also known as Khuzestan. Their backing came predominantly from Libya (where they were based) and from Iraq, and their funds had been ploughed into weaponry – the group were armed with Browning pistols, Skorpion sub-machine guns and hand grenades. The chosen target was the Iranian Embassy, hoping that an effective and highly public hostage situation would give their cause some international leverage (their demands were specifically the release of certain Arab prisoners from Iranian jails).

The siege began crudely. A British Police Constable - Trevor Lock - who was stationed on the door, found himself in a wrestling contest with the terrorists as they panicked in their attempt to enter the embassy. PC Lock tried to shut the door on them but, following shots fired through the door glass, Lock was overcome. Though a few personnel in the building managed to escape in those early,

ENTERING THE LIMELIGHT – PRINCES GATE

Since the moment of its formation in the early 1940s, much of the SAS's operational life has been spent in relative obscurity. Military historians and enthusiasts were aware of the Special Air Service because of operational contributions in theatres chaotic moments of the siege, the terrorists managed to complete the first stage of their action with 26 hostages taken. The die was cast.

At this point the events fork out in two directions. The most visible activity was the standard hostage-situation procedure of the British police. Hostage negotiators were trying to defuse the situation and buy more time, and units of police marksmen and anti-terrorist officers were overseeing more aggressive options. But, behind the scenes, the SAS was already in preparation. A counter-terrorist Special Projects Team (SPT) – known as 'B' Squadron – was given orders from the British government to go to standby in case of violent escalation. When the orders came the SAS was already in London, having driven there from the Hereford HQ to gather intelligence should they be required. This was a possibility, as Prime Minister Margaret Thatcher and her Cabinet were taking a tough line with the hostage situation, and Iran had actually turned over the handling of the incident entirely to the British authorities, something the terrorists did not expect.

The remaining terrorists now occupied the telex room and had started to shoot hostages in cold blood with sprays of pistol and sub-machine gun fire

While the siege played itself out on Princes Gate, the SAS went into exhaustive preparations. Intelligence had been provided by covert surveillance teams, who through surreptitious planting of monitoring devices could even track the movements and voices of the terrorists within the building. (This in itself was a superb operation – even aircraft flying into Heathrow were diverted to a lower flight path so that the noise of their engines would disguise the sound of drilling.) Two SAS teams – 'Red Team' and 'Blue Team' – 25 men each, were developing assault plans with the rigorous attention to detail for which the SAS is known. A full-size hessian mock up of the embassy interior had been built by the Army in Regent's Park barracks, in which the SAS troopers were testing assault tactics. Weapons and kit were obsessively

checked for reliability. SPT units are often referred to in SAS slang as 'Old Women' – they sit around waiting for the telephone to ring. There was always the possibility that – despite all the preparation – the telephone wouldn't ring and the unit would be stood down and sent back to Hereford.

But the call did come, precipitated by the terrorists, who decided to up the ante. On 5 May at around 1900 hours Abbas Lavasani, a senior embassy press officer, was shot dead and his body dumped outside the front door for collection by the police. This callous action was the watershed in the siege. The police, with Cabinet sanction, formally handed over control of the siege to the SAS at 1907 hours (the actual order was given on a rough, hand-written note, such was the immediacy of the situation). Under the command of Lieutenant-Colonel Michael Rose, Operation 'Nimrod' was launched.

Nimrod was a stunning demonstration of an elite counter-terrorist and hostage rescue operation in an urban setting. Tactically, the operation had a basic structure with three entry points into the building. Eight men of 'Red Team' were to break into the embassy from the second-floor balcony at the rear of the building (after abseiling down from the roof), with two other groups attacking third and fourth floors (the fourth by blasting through a skylight). 'Blue Team' would attack the basement and the two floors above. As the SAS moved into their positions, the terrorist leader, called Oan, signalled his mounting nerves about noises around him. There was no going back – the SAS had the green light.

What happened next was a startling demonstration of tactical excellence meeting outright ruthlessness. The SAS troopers punched their way into the building using sledgehammers, frame charges (explosive devices used for precision blast cutting) and shotguns. Once inside, they systematically cleared every room. On the first floor, Blue Team soldiers pursued two terrorists – including the leader, Oan, who had been physically tackled by the courageous PC Lock. Oan was shot on the spot, and the other terrorist was gunned down in the entrance to the Ambassador's office. On the second floor, Red Team were also rolling up the rooms with speed and aggression. A violent and explosive entry into the back office on the floor found that the terrorists had actually moved the hostages into the telex room on the opposite side of the building. Here the operation reached its most critical juncture. The remaining four terrorists now occupied the telex room and had started to shoot hostages in cold blood with sprays of pistol and submachine gun fire. One terrorist, foolishly looking out of a window, was immediately killed with a confident head-shot from a SAS sniper sited in Hyde Park. Then the SAS team burst into the room. In panic the other terrorists mingled themselves in with the hostages. The situation teetered on utter confusion. Distressed hostages were screaming in fear. The SAS quickly prioritized, and focused on getting the hostages out, evacuating everyone towards the first floor. During this process, the terrorists started to reveal themselves. One was spotted and riddled with fire as he attempted to deploy a hand grenade. Another was shot as he descended the stairs with the hostages; again, he clutched a hand grenade.

The operation was nearing its end. All hostages and the one surviving terrorist were evacuated onto the lawn of the embassy, where all were identified. The terrorist was separated (he would have been shot were it not for a hostage identifying him as one of the more benign members of the terrorist group) and put up no further resistance – life imprisonment awaited him. Two hostages had died in total – Abbas Lavasani and one in the telex room – and two had been injured, but these losses were not excessive considering the firepower available to the terrorists. Such was the speed and confusion imposed upon the terrorists (plus the fact that they had been psychologically worn down beforehand by skilled negotiators), that they were beyond rational behaviour.

Once the final terrorist had been identified, the siege at Princes Gate drew to a close, an undoubted and resounding success for the SAS regiment. Yet there was a crucial legacy of the raid which affected the entire future of the Regiment and also the nature of those who applied for SAS service. Unlike

RIGHT: The moment when SAS soldiers enter the Iranian Embassy to break the siege. Images such as these were a revelation. Never before had the public witnessed a special forces unit in action in such stark clarity. Such images would irrevocably change the profile of the SAS.

RIGHT: A famous image of the early SAS, seen here in the deserts of North Africa in January 1943 after a deep insertion mission. US Willys Jeeps provided excellent operational mobility for the SAS with the firepower of three Lewis guns and a Vickers 'K' machine gun per vehicle.

ABOVE: Lieutenant-Colonel David Stirling, the founder of the SAS. Stirling was a good judge of character, a talent which led him to pick truly capable men with which to build the SAS and establish its credibility. He died in 1990.

the actual action. This intention was thwarted by an enterprising ITN crew who positioned their camera in a flat overlooking the Iranian Embassy itself.

The result was some of the most famous combat footage of all time. Glued to the TV screen as a ten-year-old boy, I vividly recall one image – a black-clad SAS trooper leaping across a balcony division just before a frame charge detonated with a huge blast against a wall behind him. If I was one of the many children around the world hypnotized by the unfolding events, there was an equal number of men who were equally inspired. In the aftermath of the operation, applications to the SAS rocketed in number, as one SAS soldier later commented:

'Princes Gate was a turning point. It demonstrated to the powers that be what the Regiment could do and just what an asset the country had, but it also brought a problem we wished to avoid: the media spotlight. In addition, for the first few years after the siege, Selection courses were packed with what seemed like every man in the British Army wanting to join the SAS, and so we had to introduce extra physicals on the first day just to get rid of the wasters.'

Because the SAS became such household heroes after Princes Gate, many men felt that joining the Regiment would bring them a vaunted social status. Unfortunately for many of them, they were entirely missing the point. Few had genuinely contemplated what it took to get into the SAS, and many had not considered that elite regiments demand a secrecy which is at odds with a desire for public recognition. Princes Gate, the most visible face of the SAS, had ironically attracted just the sort of people the Special Air Service did not want.

EVERY MAN AN INDIVIDUAL

So what type of people does the SAS want? How should they behave, what sort of military record should they have, and how should they differ from

any other SAS action, the media was there in force. This had actually been encouraged by the British government. If there was to be a substantial and bloody military action in the heart of London then the government wanted the public to understand how grave the situation was and that SAS intervention was warranted. Yet while hundreds of journalists and camera crew gathered around Princes Gate, the intention was to keep the cameras away from

ABOVE: Parachute training for the SAS at Kabrit, Egypt, to prepare for the first operation. Stirling was initially fixed on the idea of parachute deployment as primary SAS insertion method, but navigational problems led to his adoption of vehicular methods.

the common run of men? Perhaps one of the best yardsticks for defining the ideal SAS soldier is to look back at the qualities of those who founded the regiment. By studying the originators, we see the SAS soldier at a time when the SAS had no social sta-

tus or public position, and thus was defined purely by the men who belonged to it rather than the legends it later garnered.

Who better to look at first than the founder of the SAS itself, Lieutenant David Stirling. Though Stirling is the SAS's official founder, the concept of a small unit, highly trained as a raiding party, had been conceived earlier during World War II by a Lieutenant-Colonel Dudley Clark. He had persuaded Winston Churchill and his military advisors to allow the formation of what were called Special Service Battalions, actually groups of commandos with multiple combat and operational specialities. One group of these, called No 2 Commando, was parachute-trained and thus went on to receive the name No 11 Special Air Service Battalion. They were first bloodied during an action against an aqueduct over the River Tragino in Italy in February 1941. Following the success of the German airborne invasion of Crete in May 1941, the Battalion was reformulated as part of a fledgling parachute brigade.

What really galvanized the development of the true SAS Regiment we know today was the inspiration and personal tenacity of Stirling himself. Stirling was originally an officer in the Scots Guards who joined Dudley Clarke's Commandos in June 1940 and subsequently found himself in a special strike unit in North Africa entitled 'Layforce'. During parachute training, Stirling suffered a back injury. His subsequent two-month convalescence allowed him to ruminate over a new type of military unit.

EARLY EXPERIMENTS

Stirling had always been a rather distracted and unsettled man, experimenting in his early years in everything from painting to mountaineering. His seeming lack of focus masked a mind that raced with ideas. The war seemed to channel his energies, however, and in his hospital bed he conceived of what would become the most famous fighting force in the world. What he envisaged was highly trained, highly mobile, heavily armed raiding parties consisting of uniquely small units. Previously, Allied raiding actions had generally been conducted by groups numbering around 200, with all the logistical sluggishness that this entailed. Layforce itself, disbanded by the time of Stirling's ruminations, was 1600 men strong. Stirling later explained that he felt a force 'one-twentieth the size' could have an equivalent, if not greater impact in the theatre. Stirling's belief was that small groups, numbering little more than single figures and rapidly deployed by parachute, would be able to target high priority enemy positions, hit them hard and with great speed, and then evacuate before the intense shock of the attack had even worn off. The actions Stirling had in mind were not concerned with the capture of territory from the enemy; rather, they were concerned with hitting the enemy a vigorous and significant blow before disappearing back into obscurity.

His most immediate obstacle, however, to giving life to this plan was the British Army chain of command. The commander of the North African region at the time was General Sir Claude Auchinleck, not an easy man with whom to get an audience. After Stirling's attempt to deliver his proposal in person to Auchinleck at Middle East Command HQ met with official obstructions, he tired of waiting and actually climbed the security fence and made his way towards Auchinleck's office. He was stopped on the way by the Chief of Staff, Major-General Neil Ritchie, who provided the first official ear to Stirling's plans. Ritchie was impressed, and forwarded the plans on to Auchinleck. The ideas struck a chord with the British top brass and they granted Stirling a force of 60 men, and also gave him a promotion to Captain. In July 1941, L Detachment Special Air Service Brigade was born.

For what was to become such an elite force, the early days of the SAS were incredibly ad hoc and experimental. There was little structural and financial support for the new unit from the wider British Army, and there were few models for Stirling and his men to follow. Improvization became the keyword, but Stirling understood that each man had to be a master of basic combat skills if he was to survive in the expected combat role. Consequently each man was exhaustively trained, far beyond the standards usually given to a British Army soldier. Whereas a regular army man would learn the skills to fire the Lee Enfield, the Bren machine gun, and the Sten sub-machine gun, the SAS soldier would have supplementary lessons in all the firearms of the Axis powers, should he need to use them (the SAS today follow the same principles of firearms training). Instead of overt and conspicuous combat tactics, the SAS man was taught to apply surprise, ambush, ruthlessness, and stealth. Desert survival was also taught, and navigational skills were paramount. What emerged at the end of training was a force that was truly unique – in outlook, talents, breadth of training, and endurance.

Yet despite the evident proficiency of Stirling's men, the first operation was a costly disaster.

Because of his original conception of what the SAS would be, Stirling was wedded to the tactical principles of parachute deployment, and this was the central problem behind the operational maladies which struck the young unit during those first actions. One of the primary targets for the SAS was Axis airfields. The flat, open spaces of North Africa made ground-attack aircraft an acute problem for the logistical activity of both sides. If Axis airfields could be hit continually by the SAS then the Allies could inch towards localized air supremacy.

On 16 November 1941 the SAS's first operation was launched. The targets were a series of Axis airfields around Gazala and Tmimi. The plan was for the SAS to be parachuted in 12-man squads about 32km (20 miles) from each airbase. They would advance and attack the targets, use demolitions to destroy aircraft on the ground, and then retreat to be picked up by vehicles from the Long Range Desert Group (LRDG).

The raid began in the worst possible circumstances. The weather was atrocious and it dispersed

Army's top brass at the time, who resisted disbanding the new-born Regiment). Undaunted, he immediately set about reformulating his strategies for both insertion and attack. SAS soldiers – it must be remembered by all those who intend to join the Regiment – are tough but they are not superhuman. However, they are distinguished by an irrepressible skill at solving problems. Stirling could have folded

The SAS followed an audacious strategy of simply driving down the enemy runway at speed while spraying the stationary aircraft with machine gun fire and hurling grenades

up with embarrassment; instead he abandoned the unpredictabilities of parachute deployment in favour of close-insertion by LRDG teams themselves.

This proved itself to be the winning formula. Between 8 December 1941 and 26 December 1941, L Detachment visited eight Axis airfields and netted 90 enemy aircraft destroyed. Most were taken out by hand, with the SAS troopers going right into the heart of the airfield and planting Lewes bombs – special explosive devices designed to rip open aircraft wings and ignite the fuel inside – directly onto the aircraft. Sometimes the operations would be conducted under blistering firefights in which the SAS produced heavy, but exceptionally controlled, fire. Other times the enemy would not even be aware of the presence of the SAS until the aircraft or vehicles blew to pieces. In 1942 the operations took on an added dimension when Stirling acquired some US Willys Jeeps which he converted to take Vickers and Browning machine guns plus copious jerry cans of petrol. This led to an audacious strategy of simply driving down the enemy runway at speed while spraying the stationary aircraft with machine gun fire and hurling grenades. (The strategy

both the transport aircraft (the pilots had had little training in parachute deployment and were inexperienced with the weather conditions) and the parachuting soldiers. Navigation on the ground became next to impossible. The operation was effectively called off, and the SAS troopers began an arduous and embarrassing retreat to safety. A mere 22 of the 62 men actually made it back.

Nevertheless, Stirling's response to this disaster is an excellent example of the SAS mentality (and indeed a glowing recommendation of the British

actually came about by accident during a raid on the Bagoush airfield on 7 July 1942. Many of the Lewes bombs did not explode so, leading the way, Stirling simply took his jeep onto the airfield.) The results speak for themselves – 113 Axis aircraft destroyed during the month of July alone.

By now the SAS was starting to grow in notoriety. German and Italian airfield personnel lived in a constant fear – well founded – of being attacked. For its size of unit, the SAS was having a disproportionate effect on the Axis war machine in North Africa. Stirling ascended the ranks, becoming a major in early 1942. His SAS career was eventually cut short in 1943. German forces were so alarmed at the havoc wrought by the SAS that they had formed specialist units dedicated to counter-SAS actions. In January 1943 one such unit managed to capture Stirling. He was to show all the courage and resistance he had in action, escaping three times from German captivity before being consigned behind the secure walls of Colditz Castle.

What manner of man?

By the time of his capture, Stirling had reached something of a legendary status amongst the British Army, particularly in Africa. His capture did not leave the SAS bereft of leadership or personal inspiration, although his absence did require some acclimatization. There can be no such thing as a weak link in the SAS – everyone must be the best in their trade so that mission performance does not hang upon a few key individuals. Exceptional men such as John 'Jock' Lewes and 'Paddy' Mayne stepped in to fill his shoes. These two characters became legendary within SAS history for their tenacity and daring, and their seeming invulnerability to indecision and fear. They guided the Special Air Service forward to greater and greater operational success from North Africa through Italy and the liberation of occupied Europe.

For us today, what are the overall lessons that emerge from personalities such as Stirling, Lewes, and Mayne? Who is suitable for the SAS? Perhaps we should first reverse the question, and ask what type of person the SAS does not want within their ranks. One of the first category of 'unsuitables' are mindlessly aggressive individuals. There is, in certain

RIGHT: General Sir Peter de la Billière, former head of 22 SAS and here seen in his role as commander of British Forces in the Gulf War. He epitomizes the broad talents of SAS officers, arriving in the Gulf fluent in Arabic and with a deep knowledge of Arabic culture.

quarters (usually civilian), a skewed belief that to be in the Special Air Service you have to show a capacity to be more violent than the general run of military personnel. (The extreme of this perception was a tragic case in the UK, in which two teenage boys murdered a taxi driver in the misguided belief that you had to have killed someone to enter the SAS.) While it is true that the SAS soldier must be able to deliver violence with incredible ruthlessness if called upon to do so, there is a crucial distinction between crude aggression and controlled aggression. The SAS sees more than its fair share of applicants who think that coming out on top in a bar room scrap makes them 'the right stuff' for SAS membership. Ironically, such evidence will almost always preclude an individual from joining. The SAS wants people who are self-controlled, not ruled by ego or a volatile temper. Those not possessing such self-control will not only usually lack the discretion required of an SAS trooper, but they will also tend to display bad decision-making skills in complex combat situations.

Captain Derrick Harrison, an SAS commander who operated in Sicily, Italy, and France during World War II, gives an insight into the combat mentality of the SAS. On 23 August 1944, Harrison was leading a Jeep column of SAS troopers into the French village of Les Ormes, which was under attack by a German column. Upon entering the village, the unit was confronted by a sizeable SS unit, including two trucks of troops and a staff vehicle. The SAS were heavily outnumbered, but note Harrison's response and his own sensations during combat:

'I had grabbed my carbine and was now standing in the middle of the road firing at everything that moved. Germans seemed to be firing from every doorway. I felt my reactions speed up to an incredible level. It was almost as if I could see the individual bullets coming towards me as I ducked and weaved to avoid them. And all the time I was shooting from the hip, and shooting accurately.'

The picture we get from Captain Harrison's description is not of a man exploding with violence. Instead he is a model of controlled response, becoming acutely aware of his surroundings and directing his fire with purpose and decision despite there evidently being a huge rush of adrenaline. This is the type of fighting mind the SAS is looking for. If we were to sum it up, we might say that the SAS wants men who are capable of directing violence with purpose and control rather than succumbing to the fog of war that commonly afflicts their enemies.

All this brings us onto the next essential quality of the SAS trooper – intelligence. Strangely, this can often be overlooked by applicants, as cerebral qualities are often overshadowed by the media's focus on the dynamic physical action of the SAS in combat. If this is the perception then the applicant will have a completely incorrect view about what life in the SAS is all about. SAS combat actions are usually the end of long processes of intelligence gathering, tactical assessments, practical planning, and intense liaison with other branches of the armed forces. It is never the case that the SAS soldier simply grabs his M16 or MP5, slams in a fresh magazine, then goes charging into action without any thought or strategic planning.

What must be understood by the applicant is that the SAS generally requires higher IQs than is usual for military service. This is because the operational remit of an SAS soldier is that much greater than a regular infantryman, as is the demand of self-reliant decision-making. The memorable Princes Gate action is a case in point. The media, and popular imagination, focused on the moment of combat itself, the drama of abseil assaults and the intense excitement of the room-by-room combat. However, if you examine the events more closely, you see that the fight itself was the culmination of processes that required a long and deliberate application of intelligence. For a start, from the moment they were alerted to the developing situation on 1 May, the

LEFT: IRA terrorists open fire with pistols on the streets of Northern Ireland. For over 20 years, the IRA were one of the SAS's main targets. Such actions as Loughall, 1987 – when the SAS killed eight IRA members launching a bomb attack – dented the IRA's operational efficacy.

troopers instantly had to design an Immediate Action (IA) option – a plan that could be rapidly implemented should building entry be required at very short notice. With this in place, they then had to construct a more advanced and pre-planned assault. Factors in the equation included the following: terrorist weapons available, routes of movement around the building, the physical construction of doors, windows, and wall partitions, the condition of the hostages and how they were gathered, the atmosphere that would be generated once the assault was underway, routes of entry and exit, possible terrorist responses, correct selection of weaponry, and so on. A huge amount is expected of SAS soldiers (sometimes too much) and not every person is prepared to take on such responsibility. The pressures of the Princes Gate assault are made evident by the comments of one of

The Regiment is keen to impress that SAS warfare is not about learning a few classified combat techniques and then rushing out to the world's hotspots

the troopers involved in Operation Nimrod, speaking about the moment he arrived at the embassy and was given his instructions:

'When we had arrived at the start of the siege, we had been told to be ready to storm the building within 15 minutes. This would mean going in using firearms, stun grenades and CS gas and trying to reach the hostages before they were killed. At that stage we had no idea of the hostages' whereabouts. I looked at the embassy and thought of clearing 50 rooms one by one, while all the time looking for the terrorists and their prisoners.'

As it turned out this Immediate Action solution was not tried out, but even after considered planning the burdens on each soldier were great, if not even greater.

Not only did each SAS soldier have to engage in complex tactical planning, but he even had to complete tasks such as memorizing the faces of the

LIEUTENANT JOCK LEWES

Lieutenant Jock Lewes was a Layforce No. 8 Commando member, and became one of L Detachment's first troopers. His significance to the history of the SAS lay not so much in his fighting example – though he went into action during the early airfield raids – as the direction he imparted to SAS training (he was also the inventor of the Lewes bomb). Though Lewes's passion for the parachute was initially equal to that of his friend, David Stirling, he went on to help establish a training programme for the SAS recruits that set them apart from the rest of the military world. Johnny Cooper, another original of L Detachment, described the end result of Lewes's training programme in terms of a total sense of independence and self-reliance, a resistance to discomfort, plus an unusually high level of ability in soldierly skills such as navigation and weapons handling. The SAS today is indebted to Jock Lewes because his vision of the self-disciplined, utterly competent SAS soldier is essentially the same today. Sadly, Jock Lewes' period with the SAS was all too short. On 26 December 1941 he was killed after being hit by a cannon round from an Italian aircraft making a strafing run during the SAS attack on Nofilia airfield, of which Lewes was the leader.

hostages from photographs – and translate that into split-second recognition in combat – as well as have the discipline to stay within his limited role once the action was underway.

The distinctive intelligence demands placed upon an SAS soldier are also evident by what happened after the raid was successfully completed. After bagging up their weapons for police evidence, the troopers who took part in the raid had to return to Hereford and face some 36 hours of debriefing questions from the Metropolitan Police for the forthcoming inquest. Then evidence had to be given at inquest (mainly by the command elements of the raid). Not all public bodies are sympathetic to the SAS, and the troopers' actions were examined for their legality and questioned over the appropriateness of the force used. Such questioning has a farcical edge to it – the controlled and reflective atmosphere of a courtroom bears no resemblance to the split-second decisions of combat – but the SAS trooper has to have the mental acuity to apply legal restraints on his actions and be prepared to justify them in a court.

An example from recent history is the inquest into the killing of three Irish Republican Army (IRA) terrorists by the SAS on Gibraltar in March 1988. The three terrorists were killed by an SAS team as they prepared a car-bombing against British troops on the island. While the intention of the IRA members has never been questioned, the fact that when killed they were all unarmed caused great international controversy. The SAS soldiers involved in the incident had to give evidence about their actions to justify such extreme measures. Although the court finally accepted that the killings were justified – the SAS troopers believed that the terrorists were about to pull out concealed firearms or detonate a bomb using a remote trigger – this incident certainly shows how the modern-day SAS soldier must inevitably accept the extra demands for accountability that come with his job.

Such demands are often overlooked when soldiers apply for the SAS. But they have to be recognized. The Regiment is keen to impress that SAS warfare is not about learning a few classified combat techniques and then rushing out to the world's hotspots. Rather, an SAS trooper is required to show a maturity of intelligence and practical thinking that will not only serve him well in action itself, but will also enable him to handle the planning, organization and, possibly, legal demands that his job will place upon him.

Furthermore, the SAS has need for well-rounded individuals who can also execute the many non-combat roles that may crop up within their remit. 'Hearts and minds' operations have been commonplace in the SAS from the moment the term was coined by Sir Gerald Templer during the Malayan Emergency in 1952. During that conflict, SAS soldiers lived amongst native peoples and became intimate with their different ways of living, traditions, customs, and language. (Linguistic skills are still held in very high regard in the SAS, and many soldiers will undergo training at the British Army Language School.) In return the troopers delivered an amazing variety of support to the locals. They

ABOVE: SAS soldiers have to exhibit flexibility, as their operations are often built around 'hearts and minds' actions aimed at securing the support of indigenous peoples. Here SAS troopers are seen living in a native hut in a village in Borneo, gaining the trust of the locals.

helped construct safe water supplies, build forts and civilian dwellings, as well as giving medical treatment (including midwifery services). They even learnt techniques of animal husbandry and basic veterinary care.

The hearts and minds aspects of SAS operations continue to this day, and should be considered by any applicant who is looking to the SAS for a life of endless combat. SAS operatives must show an unusual versatility of mind, as well as the ability to understand other people's perspectives and empathize with their concerns. Bullish individuals who simply want to express power over others are obviously not suitable for this sort of role, and thus they tend to be screened out at an early stage of the SAS recruiting process.

THE PROBLEM OF EGO

This brings us to an issue we have already touched upon – status and ego. There is no doubt about it, belonging to the SAS is the most prestigious military vocation in the world. Being in the SAS will bring you respect and acknowledgement from your

Above: Medical skills are one of the SAS's most useful 'hearts and minds' qualities, especially when in developing countries. Here an SAS trooper treats a native of Borneo. Other medical activities may include delivering vaccination programmes, and even midwifery.

former colleagues, and will put you at the very top of the military ladder of expertise. Yet if this status, this recognition, is the primary reason why you attempt to join the SAS, then the Regiment is unlikely to accept you.

For a start, the Special Air Service is an organization founded on a very understandable secrecy. Only with publications emerging following the Gulf War did the world of the SAS really open up to out-side view, though even this has not revealed all that occurs within. Despite the fact that many of its training techniques and even operational details have been revealed in one form or another, the

Regiment still likes to deflect outside interest. Much of the reason is simply that elite soldiers value anonymity. If lots of friends and associates outside of the Regiment know that you are SAS, it is not too difficult for a talented enemy or terrorist agent or intelligence officer to find out the details himself. Once anonymity is blown then the agent and his family are vulnerable. (This is particularly true in the case of the lasting enmity between the SAS and the IRA. SAS operatives involved in key actions against the IRA are likely to remain potential targets for many years to come, regardless of the political situation in Northern Ireland.)

So public recognition and personal glory is not a welcome by-product of SAS membership. SAS soldiers tend to join the Regiment because they have a desire to excel in the practice of men-at-arms – that excellence should be its own motivation. This is the only attitude that will sustain an SAS trooper through some of the lonelier parts of his work. An SAS sniper, for instance, may have to sit for days on end waiting to deliver a shot that no one else will witness. He must be entirely self-reliant in the process, needing no external praise, approval, or guidance to operate. This ability is not present in everyone. Many soldiers, quite rightly, need the support and social system offered by large scale units of close friends. The isolation accompanying some elite missions makes such a person prey to fears, boredom, and decision-making insecurity.

Moreover, there is a reverse to the habit of self-reliance. The SAS is a unit which has to be deeply dependent upon the resources of the broader British forces. For example, deployments by land, air and sea are usually conducted in tandem with units from the Army, Air Force or Navy respectively. During the Gulf War, RAF helicopter pilots would deploy the SAS into the deserts of Iraq (at great personal risk to themselves), while the UN air forces would act on SAS instructions to deploy air strikes. During the Princes Gate action, the SAS soldiers' planning would have been considerably more difficult had not an Irish Guards Pioneer unit accurately and quickly constructed a mock-up of the interior of the Iranian Embassy.

The point is that the SAS needs to be on very good terms with the rest of the military world if it

LIEUTENANT COLONEL 'PADDY' MAYNE

Lieutenant-Colonel 'Paddy' Blair Mayne's prodigious career with the SAS lasted until 1945. Like Lewes, Mayne was with L Detachment from the start in its initial recruitment batch, and was always to be in the thick of the action. Under fire he was cool headed, but had a spirited attitude to danger which led him to demonstrate bravery and courage bordering on the excessive, though he would never put other men's lives in danger. Mayne was a formative influence behind the SAS's switch to Jeeps for desert operations rather than relying on the LRDG. In combat, Mayne personally destroyed over 100 enemy aircraft and his military career in World War II is littered with examples of incredible, even single-handed, actions against German military forces. Mayne became commander of 1 SAS after Stirling's capture on 27 January 1943, at which point he received his promotion to Lieutenant Colonel. Later, when 1 SAS was split into two elements – the Special Raiding Squadron (SRS) and Special Boat Section (SBS) – he took command of the SRS throughout the Italian campaign, before leading the redesignated 1 SAS on the Second Front in France, the Low Countries, northern Europe, and Germany itelf. His wartime service yielded him the DSO and three bars.

is to function effectively. Hence the attitude to outside units is never dismissive, and the SAS is always the first to admit that other regiments have capabilities as good as their own, if not better, in certain areas (the big difference is that the SAS trooper tends to come as a complete package of skills). There is, admittedly, a Regimental pride within the SAS and a positive dislike for many non-combatant positions (lovingly referred to as REMFs – 'Rear Echelon Mother Fuckers'), but the SAS wants people who will cooperate with the wider military, not look down upon it.

TOUGHNESS AS STANDARD

If we consider the qualities of the archetypal SAS member, there is the problem that elite units tend to value individualism over conformity. Remember Stirling – here was a man who broke into a British

Army camp and headquarters in an attempt to promote his vision for special forces. Thankfully someone was present inside the HQ to receive the idea – Stirling could just have well have been charged. Stirling was a truly original thinker, someone who would never accept someone else's version of events and who was constantly looking out for more competent approaches to military problems. His tendency during tactical instruction lessons was to be slightly defiant, and he made enemies of senior officers as he ridiculed text-book tactics.

Such defiance is not warranted from those wanting to be in the SAS, but a passion to think for oneself and for overcoming problems is essential. One point readily apparent in reading the biographies, autobiographies or operational accounts of people such as Stirling, de la Billière, Andy McNab and other legendary SAS troopers, is that they were or are all

The instructors are looking for many different qualities... but the dominant trait of character they want is someone who will never give in to adversity no matter how hard it bites

men with a capacity to push themselves relentlessly in pursuit of a successful mission. Even against fatigue, injury, and the prospect of violent death, such men seemed to retain a 'switched on' attitude which never lapsed into defeatism and fatalism.

This is perhaps the most vital area of the SAS 'personality' – mental toughness, before its physical counterpart. For people with good levels of discipline and commitment, physical strength up to SAS levels can be acquired by many. All it takes is some months of diligent training. However, mental resilience and a determination not to be beaten by adversity will make the body accomplish even more extraordinary feats when others have dropped by the wayside. Former SAS soldier Chris

Left: An SAS patrol photographed during the Gulf War. The image is reminiscent of the Regiment's first desert operations in North Africa in the 1940s. Here the transport is a trials bike and a Land Rover 100, armed with a Mk 19 automatic grenade launcher.

Ryan served as one of the eight-man 'Bravo Two Zero' team which was inserted into Iraq during the Gulf War. While many other members of the unit were ultimately captured and tortured by the Iraqis – the endurance of which itself required enormous amounts of mental courage – Chris Ryan managed to escape and walked for nearly 300km (186 miles) across the desert to safety. His comments about this epic journey in his book *The One That Got Away* are revealing for what they say about personal toughness in the SAS:

'Before the Gulf War, if somebody had told me I could walk nearly 300 kilometres through enemy territory in seven nights, with no food and practi-

cally no water, with inadequate clothes, no proper sleep and no shelter, I wouldn't have believed him. When I had to, I did it... In 1991 I was at a peak of physical fitness, and armed with the skills, the endurance, the competitive instinct and the motivation which SAS training had instilled in me.'
(Chris Ryan, *The One That Got Away*, London: Arrow Books, 1995)

Note that fitness is only one reason why Ryan was able to endure his incredible journey across the

FORMATION OF 22 SAS

Today's 22 SAS Regiment began its life because of the vision of one man in particular, Brigadier Mike Calvert. A former Chindit and an expert in guerrilla tactics, Calvert commanded an SAS Brigade until 1945, when the SAS was generally disbanded. However, during the Malayan Emergency in the 1950s, Calvert was tasked with a research project to find out the best way of combating the social and military strengths of anti-British communist insurgents. Calvert's eventual recommendations led to the creation, under his command, of the Malayan Scouts (SAS), an elite small-unit force designed specifically to employ jungle-based counter-insurgency tactics. The Scouts began to have a some impact on the terrorist presence, an impact which was greatly increased when reservist members of 21 SAS (an SAS unit retained after disbandment as a Territorial unit) and other elite World War II personnel joined them. Eventually, the Malayan Scouts were renamed as 22 SAS and once more the Special Air Service became part of the permanent structure of the post-war British Army. With the re-formation of the unit, Calvert actually returned to the UK where he left a lasting legacy for the Regiment by designing the Selection procedure which is the basis of training today.

desert. Equally, if not more important, were those mental qualities of 'endurance', 'competitive instinct' (it is no coincidence that many SAS troopers were, or have become, superb competitive sportsmen) and 'motivation'. As Ryan states, SAS training had established these values within him, but they also have to be already present in some measure if a person is to successfully attempt to pass SAS Selection.

The SAS Selection course is one of the toughest in the world. For every 150 men who start the Selection process, less than 15 will usually be left at the end of it. The instructors are looking for many different qualities but the dominant trait of character they want is someone who will never give in to adversity no matter how hard it bites. Many soldiers begin the SAS Selection course believing that they are that person, only to be proved wrong when several weeks of incredible physical punishment, sleep deprivation, and high-pressure decision-making start to take their toll. The men that are left are those whose inner resources carried them through.

Application and Preparation

Many potential SAS recruits who arrive at Hereford to begin training do not even manage to last the first week of the endurance exercises. In most cases either their motivation is not sufficient or they have not carried out the hard, physical preparation required for one of the world's most demanding training programmes.

It is common knowledge that the SAS Selection course is one of the most difficult military recruitment programmes in the world. Twice a year – once in summer and once in winter – about 150 recruits line up to begin the process. Typically, they are fairly seasoned soldiers in their mid- to late-twenties, usually very fit, as well as extremely apprehensive about what lays ahead. The apprehension is well justified. By the end of Selection, when the coveted beige SAS beret is handed out to those who have passed, less than 15 men will be remaining. An attrition rate of over 90 per cent ensures that only those who are superbly fit, militarily talented, and exceptionally motivated actually make it into the world's most elite regiment.

LEFT: Many recruits to the SAS come from The Parachute Regiment, whose own training programme is rigorous in the extreme. Here paras perform the 'steeplechase', a severe obstacle course which stretches over a mile and has to be performed in double-quick time.

The history of SAS Selection procedures is one of a gradual increase in formality and structure. During the Dudley-Clarke and Stirling years in World War II, Selection was distinctly informal. The commandos and early members of L Detachment were generally recruited through recommendation, personal contacts, and interview. There was no Selection course as such – the recruit embarked straight away on the operational training and if he could keep up (a big if) then he could stay within the Regiment. Such a process was entirely fitting with the ad hoc creation of the SAS during the war to meet immediate theatre demands.

It was only when the SAS was re-formed as 22 SAS in Malaya in the 1950s that a more structured Selection process was developed. At this time the SAS was based in Malaya, but conducted Selection at Brecon Lines army camp in Brecon, Wales – a destination for some of the SAS's most rigorous Selection exercises today. Because SAS operations prior to this date had required formidable powers of endurance, the new Selection regime was geared up to test recruits' powers of sustained effort. Using the most basic of equipment and army kit, the SAS recruits would have to spend large amounts of time in long-distance hill-walking exercises, combined with flawless map reading. This set the trend which continues in SAS Selection today, and by the 1980s the Selection course had reached the basic format that is currently used (though the course is updated and improved continually).

This chapter focuses on what it takes to get ready for the SAS Selection process. Rigorous preparation is vital. If you turn up on your first day at Hereford without being at your peak of physical and mental fitness, you will fall by the wayside in the first few days and become an RTU (Returned to Unit – i.e. sent back to your original regiment). The key point – and something former SAS personnel are quick to point out – is that you cannot start SAS training with the intention of 'winging it'. Surprisingly, many soldiers do arrive at Hereford with absolutely no

LEFT: Road running is the first line of preparation for SAS endurance runs. At first, run without a pack, then introduce weights and harder gradients to accustom the body. Swing the feet fairly low on road surfaces to minimize the risk of impact injuries to knees and ankles.

preparation. Almost all fail to make it. Instead, an applicant has to be utterly focused on meeting the demands of Selection at least three months, preferably six, before actually attending the course. This period is ideal to build himself into a state in which he has a real chance of fulfilling the SAS Selection criteria. If this is done, then he will arrive at Hereford at least confident that he is fit to begin.

APPLICATION

The most basic fact about eligibility for the SAS is that if you are not already a soldier, then you can't join. Civilians are able to join the Territorial SAS Regiments, though these do not have the same operational status as the regular 22 SAS. Following the Princes Gate action, there was an explosion of civilian interest in entering the Regiment, with lots of hopefuls travelling down to Hereford and loitering near the SAS HQ with the intention of somehow getting through the gates and entering the elite. However, the SAS accepts only British Army or Commonwealth soldiers who have already completed training with regular units. The reason is obvious – the SAS focuses on producing elite soldiers, so they do not want to have to teach their recruits the basics of soldiering. These should have already been acquired in the soldier's regular unit.

This criterion is one of the few formal restrictions of SAS eligibility. You must have a good record with your regular unit (i.e. have committed no offences), as it is up to your commanding officer to approve your request to apply for SAS training (the SAS only takes volunteers, there is no process of enlistment). A medical must be passed satisfactorily to obtain a Fit for Course Certificate (the medical is taken with your regimental medical officer). In addition, the recruitment criteria usually require applicants to have three years of military service left with their regular unit. This is for the simple reason that they do not want to train up someone who then has to leave after only a few months, though the length of service can be extended upon request. However, the criteria for length of service can alter depending upon the SAS's recruiting situation at a particular moment in time, and your regiment will be able to give you specific details when you approach about application.

Finally, before you put in your application make sure that your life is truly suited to being in the Special Forces. If you are married with children, you must think hard and evaluate whether the long periods away from home, often isolated from family contact, will damage your family life more than you can accept. Ask yourself whether you are prepared to leave behind the mates and social life that you have built up within your regular unit.

Perhaps most importantly, do you recognize in yourself a total drive and dedication to entering the Special Forces? Any form of weakness in your own motivation and drive will almost certainly mean that you will fail the Selection course, and that would be a senseless waste of your own time as well as the time of the Special Air Service and your regular unit.

If, however, you feel it is right to go ahead and your application is accepted, then the serious work must begin. There is a weekend course – the Special Forces Briefing Course (SFBC) – run at Hereford HQ which gives applicants an insight into SAS life prior to actually beginning Selection. We shall look at the content of this course at the beginning of the following chapter, but suffice it to say here that the NCOs and officers who run the course will emphasize the need to get yourself into superb shape before you even pass through the gates at Hereford. You will probably have time to do this; the queue for SAS Selection can be a long one and you will have to wait your turn. This may be frustrating, but channel any anxieties you feel into an effective training programme.

TOTAL BODY FITNESS

SAS soldiers have a rare level of fitness. There are athletes who can run faster, there are weight-lifters who can hoist greater weights, and there are stronger swimmers, but the SAS soldier must have a complete approach to physical strength that is matched by few. Borrowing a civilian training term, the SAS applicant must aspire to Total Body Fitness (TBF). TBF, as its full name suggests, is a state in which every aspect of the body is conditioned so that it can meet the multiple physical challenges that are experienced in the SAS. Characteristics of TBF are as follows:

HEIGHT TO WEIGHT RATIO

Excess fat has become a problem of almost epidemic proportions amongst all sectors of Western society, and has even become a problem in some sectors of the military (especially at recruit stage). In itself a high fat intake is not a cause for concern as long as the individual is doing a level of daily physical activity which burns off any surplus calories. Unfortunately, modern high-fat foods are so cheap, easily obtainable, and tasty that with sedentary lifestyles on the increase obesity is becoming a serious medical issue for most developed nations. Obesity significantly increases the chances of life-threatening conditions such as heart attack, cancer, diabetes, and stroke. However, serious obesity is rare in soldiers simply because they keep in a reasonably good state of health. Note the word 'reasonably' – even moderate amounts of excess fat can impair the efficient mobility and endurance required to complete SAS Selection.

What is vital is that your weight does not exceed the correct parameters for your height and frame, otherwise your body is having to do extra work simply transporting the excess – effort that could be diverted into the Selection process. Furthermore, obesity makes the heart and lungs work much harder than usual, sometimes to unsafe limits – hence the common and tragic cases of overweight individuals who bring on a heart attack through an over-ambitious campaign of exercise. In this chapter there is included a height to weight chart which gives the height/weight ratio you should aim for to achieve TBF. If you are attending SAS Selection, you should, however, have a slightly more sophisticated approach to fat. Endurance training requires fat very much as its fuel, so putting on a little bit of weight (though there should be no visible excess) may give you some energy reserves for the punishment that lies ahead (this is particularly true of the survival phases of training). Yet elite soldiers should aim to keep fats at 10 per cent or below of their total dietary intake.

The need for fats issues a note of caution to those who find themselves under their ideal weight. The SAS programme of Selection will require you to draw deep on your physical reserves of energy. If fat simply does not exist, then the body starts looking

IDEAL HEIGHT/WEIGHT RATIO

Height	FRAME SIZE		
	Small	Medium	Large
5ft 4in (1.62m)	118-134lb (53.5-61kg)	124-136lb (56.5-62kg)	132-148lb (60-67.5kg)
5ft 5in (1.65m)	121-129lb (55-58.5kg)	127-139lb (57.5-63kg)	135-151lb (61.5-68.5kg)
5ft 6in (1.68m)	124-133lb (56.5-60.5kg)	130-143lb (59-65kg)	138-156lb (62.5-71kg)
5ft 7in (1.70m)	128-137lb (58-62.5kg)	134-147lb (61-67kg)	142-161lb (64.5-73kg)
5ft 8in (1.72m)	132-141lb (60-64kg)	138-152lb (62.5-69kg)	147-166lb (67-75.5kg)
5ft 9in (1.75m)	136-145lb (62-66kg)	142-156lb (64.5-71kg)	151-170lb (68.5-77kg)
5ft 10in (1.78m)	140-150lb (63.5-68kg)	146-160lb (66.5-72.5)	155-174lb (70.5-79kg)
5ft 11in (1.80m)	144-154lb (65.5-70kg)	150-165lb (68-75kg)	159-179lb (72-81kg)
6ft 0in (1.83m)	148-158lb (67.5-72kg)	154-170lb (70-77.5kg)	164-184lb (74.5-83.5kg)
6ft 1in (1.85m)	152-162lb (69-73.5kg)	158-175lb (72-79.5kg)	168-189lb (76.5-86kg)
6ft 2in (1.88m)	156-167lb (71-76kg)	162-180lb (73.5-82kg)	172-194lb (78-88kg)
6ft 3in (1.90m)	160-171lb (72.5-77.5kg)	167-185lb (76-84kg)	178-199lb (81-90.5kg)
6ft 4in (1.93m)	164-175lb (74.5-79.5kg)	172-190lb (78-86.5kg)	182-203lb (82.5-92kg)

LEFT: Triceps extensions are just one exercise that can be performed in a modern well-equipped gym. The important factor is to develop all sets of muscles equally around the body, otherwise the physique will suffer from a structural weakness which may result in injury.

around for other types of body tissue to break down for energy, and serious health problems can begin. Your goal should be to put yourself into the correct height/weight category (probably towards the upper end of each band if you are going for SAS Selection) and to stay there through both dietary discipline and exercise. Ironically, this may involve eating considerably more than you are used to – whereas 2500 calories are a normal intake for an average man, an SAS soldier should be taking in about 4000-5000 calories in order to sustain his energy requirements.

HEART RATE

Height/weight ratio gives you a structural target for what you should weigh, but even if you are well within your boundaries it is, of course, far from an actual indicator of your true fitness. More telling is a test of your heart rate in times of exertion and at times of rest. The more unfit you are, the faster your heart rate will be during both rest and activity, as the heart needs to pump more to deliver the requisite amount of oxygenated blood to the muscles. Two simple calculations will give you a good overall indication of your fitness. (Note: these tests are aimed at men – women are still not allowed in the SAS even though the SAS has utilized women from other services at times).

The first straightforward exercise is to ascertain your resting heartbeat. Perform this test just after waking up or after sitting still in a chair for about an hour (make sure this is not just after eating – your heart will be working faster than usual to support digestion). With a watch in front of you, place the fingers of one hand on the inside wrist of the opposite arm about 1cm (0.4in) in from the thumb side. By moving your fingers around slightly you should be able to locate your pulse quickly. Never use your thumb to find a pulse as your thumb has its own pulse. Simply count the number of heartbeats in a 60-second period. Use the following chart then to assess your resting fitness:

ABOVE: The quad stretch. Stand on one leg and stabilize yourself with a hand against a wall. Stand up straight, then reach behind with your free arm and grip your non-supporting leg at the ankle. Draw the ankle up towards your buttocks, stretching the top of the thigh. Stay straight throughout the exercise. Repeat on the other side.

Beats per minute	Level of Fitness
30-59	Very good
60-75	Good
76-85	Average
86-99	Poor
100+	Very unfit

Beats per 30 seconds	Level of Fitness
33-36	Excellent
37-40	Good
41-43	Average
44-47	Acceptable
48-59	Poor

If you discover that you occupy the last two categories (i.e. poor or very unfit), then you obviously have a great deal of work to do to bring yourself up to SAS fitness standards, if you can accomplish it at all. Your benchmark should be around 60 beats per minute. However, before you decide to make a final judgement over your heart rate, you should perform a simple step test measurement. Dressed in light training gear for exercise, step up and down on a training bench or step about 15-26cm (6-10in) high for three minutes solid, making sure that you place both feet squarely on the bench and stand up straight before stepping down. After exactly three minutes, stop, wait for 30 seconds, then time your heart rate for just 30 seconds. The following chart will help you to analyse the results:

This test should be much more telling than the resting heart rate test as it shows how effectively the heart recovers after strenuous exercise. However, if you are contemplating the SAS Selection process then you should aim to make sure that your pulse rate is in the top two categories here, and the only way of doing that is to accustom your body to exertion through a coherent programme of exercise.

BELOW: The hamstring stretch. With your legs out in front of you bend at the waist and slide your hands as far down your legs as possible. Keep the back straight and bend gently – do not bounce yourself down further. Relax into the stretch, keep breathing, and shift down a little further when comfortable. Release yourself gently.

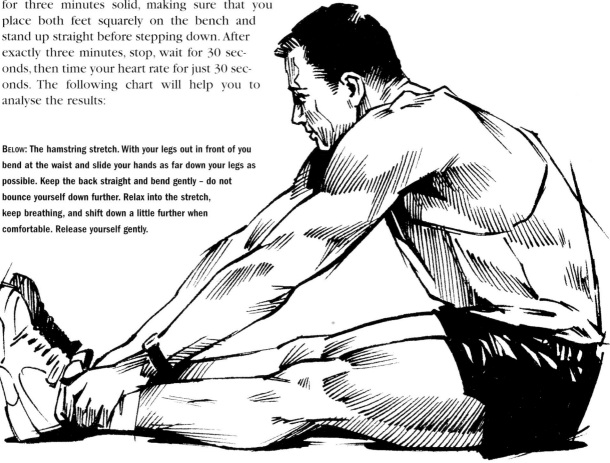

MUSCLE POWER

British army training, including that of the SAS, tends not to place as much emphasis on dedicated muscle development as forces do in the US, for example. US military training often sets a high priority on push-ups, squat-thrusts and weight training to produce powerful upper-body strength, whereas in the UK the direction is more towards endurance. Having said this, a conscious programme of strengthening all the body's main muscle groups will be vitally important to attaining Total Body Fitness. The SAS Selection process places a formidable burden on the human physique. Total kit weight on some marches can be over 36kg (80lb). Without a strong musculature, the human frame will distort under such a load and problems with breathing and mobility will ensue. Damage to the joints – particularly the knees – is also much more likely if the major muscles of the back, abdomen, and legs are not properly honed.

Total kit weight on some marches can be over 36kg (80lb). Without a strong musculature, the human frame will distort under such a load

FLEXIBILITY

Flexibility tends to be one of the most neglected elements of Total Body Fitness, particularly amongst groups such as soldiers. Flexibility is all too often equated with the type of fitness only required by gymnasts and dancers, who need their bodies to extend through an unusually wide range of movements. It is, however, a mistake to neglect this important area. Flexibility is intimately connected with body strength, as a flexible muscle allows powerful movements to be completed more efficiently than a short muscle (hence professional runners stretch diligently before and after every race). In addition, well-stretched muscles and ligaments will reduce the risk of these tissues straining or tearing during exercise, something which is extremely important for the endurance sections of SAS Selection training.

ENDURANCE

Endurance is probably the most pivotal physical quality to which SAS applicants aspire (or should aspire). The mountain-running exercises of Selection will demand that an applicant put his body through days of massive punishment which will draw out every ounce of his resources of lung capacity, muscle conditioning and stamina. Though the SAS Selection course in large measure focuses on sustained running with loads, the applicant must have a more rounded programme of endurance training. By cross-training – mixing running with swimming, cycling, circuit training, racket sports, etc., – the body will acclimatize itself to endurance in numerous different contexts. Remember, the body develops itself around the type of exercises performed. If you just concentrate on running, and then have to complete swimming tests, the endurance gained through running will not stand you in good stead as completely different muscle groups or movements will be required for the swim – you will soon find yourself exhausted. Thus Total Body Fitness requires that you are able to meet the stamina requirements of a multiple range of endurance tasks.

INJURIES

It is a sad fact that some elements of fitness are quite beyond your control and could affect whether you are accepted onto the SAS Selection course in the first place. One of the most unfortunate causes of rejection is prior injuries. As it is only soldiers who join the SAS, they will have already gone through medical screening and should be generally fit and healthy. However, if an individual incurred damage during his initial training – such as a damaged knee or ankle joint – he should be honest as to whether the injured zones will withstand the rigours of SAS Selection. Many completely fit people are ejected from Selection because they injure themselves, so those who have predispositions toward injury should perhaps reconsider their application. The best procedure for any soldier with such concerns is to visit his regimental medical officer and have a thorough evaluation (indeed this will have to be done before the applicant can proceed to the Selection process).

BELOW: Groin muscle stretch. Sitting down, place the soles of your feet together and draw your ankles in towards your groin as far as possible. Put a hand on each knee and push down gently for a few seconds. Push up with your knees against your hands for five seconds, then relax the tension, allowing your knees to be pushed down further. Avoid any burning sensations in the groin muscles.

BAD HABITS

The soldiering lifestyle is a healthy one, yet its intense social life is also conducive to developing certain habits which are inimical to Total Body Fitness. Both high alcohol consumption and smoking can impact upon your performance in SAS Selection. Smoking impairs the lungs' efficient processing of oxygen – a substance you will need in plentiful supply during the punishing endurance marches. It also makes the stamina exercises more unpleasant to do by constricting the airways. Many people who complete SAS training are smokers, but considering the long-term health risks and your desire to have every advantage possible during Selection, the preparation phase might be a good time for a soldier to quit.

Alcohol does not cause any problems as long as it is in moderation (an occasional pint can replenish energy if you are working your body hard during the day). However, excessive consumption has many dangers, including what you may do when you are drunk. A former infantryman, as a bet, drank a bottle of kettle descaling acid after an epic night of drinking. He is now officially disabled, and has to take almost all his food in puréed form because of his horribly damaged oesophagus and stomach. His military career ended instantly. Apart from the dangers of physical damage remember that senseless actions that involve the police (military or otherwise) will be reported to the SAS recruiters and can result in them precluding you from Selection.

If your drinking is heavy and repetitive (every day or every other day) then this may cause problems with controlling your weight (alcohol has a very high calorific content), and can also lead to problems with your thought processes. Selection is as mentally demanding as it is physically tough. If you have been drinking substantially the night before an exercise, your mental processes may be sluggish and crucial mistakes could be made in, say, your navigational tasks. A final bad habit is drug abuse. Drugs come in all shapes and have a variety of effects, from so-called 'soft' drugs, such as cannabis, to powerful mind-bending substances such as heroin. If you take illegal drugs, you are not right for the SAS.

Dehydration impacts on everything from energy levels to mental ability, so keeping hydrated will give you a good foundation for performance

Drugs affect temperament, thought processes, health and physical coordination, and effects can last days, even weeks, after a single usage. Moreover, the drug-testing policies of most military forces today mean that if you are caught (some drugs can be detected in the bloodstream three months after consumption), then forget SAS training – your whole military career could be over instead.

FOOD

The food that you eat is the final element of Total Body Fitness. Fitness nutrition is an advanced science all of its own, but when it comes to SAS training there are several basic facts about your diet of which you should be aware. First, eat a balanced diet. This is age-old advice, but it needs some

discipline to obey. Cut out or significantly reduce your intake of junk food and focus on eating plenty of fruit and vegetables (about five servings a day is recommended). When trying to elevate your energy levels, do not rely on sugary snacks – the energy provided is easily burnt off and may leave you feeling even more drained than before. Instead, focus on carbohydrates such as pasta, bread, fruit, pulses and starchy vegetables, such as potatoes. These will give you long-term energy that will be more useful during the exhausting Selection process.

A balanced diet will be a real asset to your training, and there are some final general points you should note to help in your pursuit of TBF. One of the most important is coordinating your eating around training. By far the most important meal of your day is breakfast. Have a large breakfast but not a greasy one – a cooked breakfast is fine but make sure the bacon is grilled, and have the eggs scrambled or poached rather than fried. Supplement the breakfast with other foods such as cereal, porridge, fruit juice, and grilled tomatoes. A good breakfast gives you a significant energy boost for the morning ahead and also contributes to a clearer mind.

Make sure that during any period of training your water intake is high – at least three to five litres a day. Dehydration impacts on everything from energy levels to mental ability, so keeping hydrated will give you a good foundation for performance. If your urine is a dark yellow colour then it means you should increase your water intake significantly; urine is clear when a person is fully hydrated.

Finally, do not eat large, heavy meals when you are in the middle of your training. Filling a hot, adrenaline-pumped body with a large meal overloads the digestion and can lead to cramps and vomiting as the tense stomach rejects the sudden load. The key principle is: eat little but often. After your substantial breakfast eat carbohydrate snacks at regular intervals (muesli bars can be good, but be careful that they are not packed with sugar). Always remember your body is like an engine – it needs fuel to run, and good quality fuel at that.

To attain all the elements of Total Body Fitness requires serious discipline. TBF may seem to have an almost monastic fanaticism about it, but our

BELOW AND OVERLEAF: You will do a lot of push-ups in SAS training, so you must do them right. Using the correct technique: the back is straight and the entire body is level throughout; going down, the arms are bent until at right angles at the elbow; going up, the arms are fully straightened.

specific context here is getting ready for SAS Selection. Remind yourself that of every 150 recruits that begin Selection, about 15 are there at the end of it. If you are going to attempt it, you need every scrap of physical advantage to stay in the game, so the diligence and comprehensiveness of TBF are justified.

However, overtraining can be as much a problem as undertraining for SAS Selection. Some individuals are so desperate to pass the Selection that they push themselves beyond healthy limits. Rest and relaxation are as important to developing fitness as the actual exercises themselves. Fitness advances by the body's muscles breaking down and then rebuilding themselves, but if exercise is relentless then the muscles do not have the chance to strengthen up. In addition, a body in a constant state of fatigue will create problems with digestion, as well as producing alarming mood swings and an inability to concentrate. At least one day a week should be kept exercise free.

BUILDING UP FITNESS

Finally, we shall now take a closer look at the forms of exercise you should be pursuing in readiness for SAS Selection. If you do not have the knowledge yourself of, for example, weight training or flexibility development, it is worthwhile putting yourself in the hands of competent and experienced trainers or instructors. This means that you will not waste valuable weeks on inadvisable or incorrectly performed exercises. With SAS Selection approaching, every hour has to be spent wisely to ensure that you are ready.

GETTING READY TO EXERCISE

Warming up is actually one of the most essential parts of your preparation regime, and it should be diligently practised before any form of exercise. It involves slightly increasing your heart rate to prepare it for exertion and also raising the temperature of your muscles. This last point is the most essential. When muscles are cold they are at their most inflexible. Suddenly plunging them into exercise can severely damage them, resulting in pulled or torn muscles or ligaments. By thoroughly warming up, the muscles are looser and more flexible and run a significantly lower risk of injury.

Warm-up exercises can be a matter of personal choice, but whatever system you follow make sure that the routine is not too vigorous and that it works systematically through the body. An example of a warm-up routine might be:

● Walk briskly for about 5 minutes, thus increasing your heart rate and temperature.
● Warm up the neck muscles by gently twisting the head from side to side, then circling the head 180 degrees with the chin moving from one shoulder, past the middle of the collarbone, to the other shoulder. Finally, rotate the head in a full circle, but do not let your head go too far backwards – stretch up rather than back.
● Swing both arms forward in large circles several times, then reverse the direction.
●To warm up the muscles of the torso, stand with

both feet shoulder-width apart and twist from the waist from side to side with your arms held up. Try to look behind you to help the exercise, but twist the head with the shoulders, do not snap it back beyond its natural limits. After this exercise, place your hands on your hips and simply rotate them as if spinning an imaginary hula-hoop. Do this about five or six times in one direction, then reverse the direction.
● The walking you did to start should have warmed-up your leg muscles and joints, but a few simple stretches will also help. Standing on one leg, take hold of the other leg just below the knee and pull the knee up to the chest as far as it will comfortably go. (Remember: stay standing straight while you are doing this; do not lean your torso towards the knee.) Repeat for the other leg. Switching back to the first leg, stretch your hand down behind your body and take hold of your ankle. Pull the leg backwards so that the back of the heel touches the back of your thigh (or as far as you can go). Repeat on the other side.

This simple routine can save you from a whole realm of muscular problems during exercise, so make it an unfailing part of your training ritual.

MUSCLE DEVELOPMENT

Muscles are long groups of strong tissue fibre attached to the skeleton and capable of contraction and expansion according to whether the person needs to pull, push, or apply leverage. When exercise is taken which impacts on the muscles, some of the fibres actually break down under the strain but then rebuild themselves with greater strength to meet the new challenge.

Controlling this process is the essence of muscle development. Through weight training and other muscle strengthening exercises, muscle tone can be improved systematically and the load-bearing strength required for SAS Selection can be achieved. However, one or two vital principles apply when approaching weight training. First, muscle strength must be developed slowly. Rushing to the gym, overloading a weight bar, and attempting to bench press far beyond what you are used to is an ideal way to inflict severe, possibly irreparable,

damage on muscles and joints. Instead, always attempt what is comfortable first and work from there (the biggest danger is ego – do not try to keep up with those around you in the gym, set your own routine). Next, for TBF to be complete, every muscle group must be developed. Muscle groups are often interconnected in their influence. Strengthen one but not its immediate neighbour, and the result can be structural insecurity. For example, many people engaged in weight training focus on developing their abdominals (often because sharp 'abs' look good) while neglecting the erectors (lower back) and latissimus dorsi (middle back). If this is done then the trainee might find an increase in lower back pain owing to a disproportion in the solidity of the torso. Finally, remember that SAS soldiers look incredibly well honed but they never look like body builders. Over-developed muscles are more of a hindrance than a help, since they often contain a high percentage of scar tissue (damaged muscle fibres which are inflexible and have no effective strength) and, if bulked up too much, can actually restrict movement. The best guide for muscle strength is simply that your body is strong enough to do the tasks which you are setting it.

There are two basic forms of muscle development exercise – without weights and with weights. Training without weights is convenient and effective, and has the advantage of your being able to do it almost anywhere. Most people are familiar with the following exercises, though there are some points you should bear in mind when training:

PRESS-UPS

When practising press-ups, go through the full range of movement – from arms locked to nose near the ground. Also keep your body rigid and back straight (do not tilt your hips upwards so that you

LEFT AND ABOVE: Stomach crunches are ideal for building up abdominal strength. Lie flat on the floor with the hands behind the head. Then curl up, bringing your knees in and drawing your upper back off the floor using the abdominal muscles. Then uncurl to the start position. Do not pull on the neck with the hands and keep the small of the back on the floor.
FAR LEFT: Press-ups test both stamina and strength – Special Forces soldiers must be able to do at least 42 full press-ups in two minutes.

ABOVE: Dumb-bell weights build up the strength of shoulders, upper back and arm muscles. Sit straight with legs in a wide supportive position. Hold the dumb-bells out to the side with arms at a right-angle at the elbow. Then push the weights straight up until they touch above the head. Lower to the start position then repeat.

are in a shallow A shape). Do not be tempted to practise rapid, shallow press-ups just to clock up the numbers, for the SAS DS (Drill Sergeant) will not accept any short cuts. It is amazing how different – and exhausting – properly performed press-ups are when compared with hasty press-ups, so make sure that you perform them properly well in advance of Selection. For added muscle development, try to vary the positioning of your hands. Sometimes have them spread out very wide, and at other times place them practically on top of each other. This will have the benefit of isolating different muscle groups within the arms and shoulders.

CHIN-UPS

Chin-ups require a bar to be performed, and are an excellent method of developing the arms, shoulders, chest, and upper back. Like press-ups, they must be performed properly to maximize the results. Use a good wide grip and when you pull up to the bar take the bar to throat level. Cross the legs

Do not just aim to lift very heavy weights with low repetitions. Heavy weights will build up muscle mass, but can also make the muscles very slow on the move

at the ankles. Do the chin-ups fairly steadily and without jerking, and concentrate as much on a steady release as the pull-up.

PARALLEL DIPS

Parallel dips are best performed on the correct equipment – two parallel bars. However, if these are not available to you, then a simple variant exercise can be carried out between two immovable pieces of furniture, with your heels on the floor extended in front of you (but taking none of your body weight). In both cases the principle is exactly the same for this exercise: you take your full body weight on straight arms between the two poles, then lower yourself gradually to the point where your upper arms are parallel to the floor. Then steadily push yourself up again. Parallel dips are excellent for producing strong biceps, triceps, and shoulders.

CRUNCHES

Crunches are a much more productive exercise for developing abdominal strength than the traditional sit-up (though you should practise sit-ups as you will probably face them during SAS Selection). This is because sit-ups rely to a large extent on the back muscles to power the torso into an upright position, rather than purely isolating the abdominals. Crunches differ in that the torso is just curled up enough to place the abdominals under tension. One technique is to have the feet flat on the floor with the knees bent up, but lifting the knees to the chest to meet each curl will work the lower abdominal muscles too. Whichever you perform make sure that your back is not under severe tension and that the small of your back remains on the floor.

Straight crunches tend to develop the frontal abdominals only, but more rounded abdominal development can be achieved through mixing conventional crunches with twisting crunches. In this case, as you pull up, twist the torso to one side, aiming one shoulder (your hands are placed behind your head in crunches) at the opposite knee. Then repeat for the opposite side.

LUNGES

Lunges are simply performed by standing up straight, then stepping forward with a long step on one leg and bending the knee until the thigh is parallel with the ground. Then you step back up again and repeat on the other side, and keep the alternation going. This exercise helps build up the thighs and upper legs, but focus on keeping your body posture upright when stepping forward as bending will reduce the effectiveness of the exercise.

There are many more weights-free exercises, but the important point is that you practise all through the full range of movement to maximize their impact. Training with weights offers a more advanced method of muscle building. Weight training can be conducted with free weights at home, but gyms with formal weights often have the advantage of expert trainers who can guide you through the best methods of using the equipment.

The amazing array of weights and endurance machines that you will find in a well-equipped

Above: Dumb-bell presses build up the pectoral, biceps and deltoid muscle groups. Lie on an exercise bench facing upwards. Hold the dumb-bells out to the side in extended arms, but keep the elbows bent. Then bring the weights straight up the middle above the chest. Release slowly until in the start position and then do repetitions.

weights room can be distracting and lead to a butterfly approach to strength training – hopping from one piece of equipment to another. It is far better to find a limited range of exercises for each muscle group and practise them diligently. There are also some important principles which you should follow with all weights training, so that it doesn't become more of a hindrance than a help to your Selection training.

Firstly, do not just aim to lift very heavy weights with low repetitions. Heavy weights will build up muscle mass, but can also make the muscles very slow on the move. Therefore mix the heavy weights/low repetitions with light weights/high repetitions. This allows the muscles to retain good speed responses, something you will need during later elements of SAS training. Always build up your training from weights that you can comfortably handle to something more strenuous – do not try to keep up with those around you if they are obviously stronger than you (in any case, it does not actually mean that they are fitter than you).

Secondly, divide your training session equally between four zones: upper body (shoulders and chest); arms and back; abdominals; legs. By giving the correct attention to each zone you will not end up with a structural imbalance in your musculature. Also, do not train for too long; an hour should be the maximum as you might find that you overwork the muscles by doing more. Finally, and perhaps most important, allow at least one day off between each weight session for your muscles to recover – you want to avoid arriving at SAS Selection walking and moving like an old man.

With all the different types of weight training, watch out if you suddenly feel sick, dizzy, or in unusually severe pain. If you observe these symptoms stop immediately and rest. If the symptoms persist, see a doctor.

FLEXIBILITY

Like strength training, flexibility training should be conducted slowly and gradually. When performing any stretching exercise only take a stretch to the point where the strain is felt and then stop, relax into the position for a few seconds, and take it a little further. Repeat the bend, relax, bend procedure

UNARMED COMBAT TRAINING

Because of the emphasis on weaponry in the modern armed forces, many soldiers arrive at SAS training with little unarmed combat training other than the basic techniques they learnt in the regular army. Taking up a martial art can assist with the speed, power and reflexes required to perform well in the SAS's Close Quarter Battle unarmed combat training. There are a huge variety of martial arts to choose from, and not all are appropriate. Choose a 'hard' art – training which emphasizes the overcoming of opponents by force and power (karate, kickboxing, kung fu, etc.,) – rather than a 'soft' art such as Aikido which concentrates on blending with the attack and repelling by generally non-damaging throws and pushes. Good martial arts to try specifically for SAS preparation are Ju Jitsu, Karate (particularly a close-quarters form such as Goju Ru), kickboxing and hard styles of kung fu. Alternatively, join a reputable self-defence class which teaches realistic self-defence – remember that delicate wrist-locks will not work on highly aggressive and non-compliant opponents set on killing you. SAS CBQ trains in a few very effective and straightforward techniques – it is not a martial art but some prior preparation will help your body respond quickly to the training.

as far as it is comfortable and then slowly release yourself out of the stretch. If any pain or a burning sensation is felt then you should stop immediately. Never try to impersonate the horrendous forced stretching exercises you might see in some martial arts movies – they will only result in potentially serious injury and a lack of fitness for any training course, let alone SAS Selection.

By approaching flexibility in a disciplined and progressive manner the muscles are, as in strength training, broken down but then rebuild themselves in the new elongated structure. Allow as much time as possible to build up your flexibility; it can take many months to attain a reasonable level. For SAS Selection training you should particularly work on hip flexibility, as this will help with the marching, running, assault courses, and swimming tests.

There are stretches designed for almost every muscle in the body. A good Yoga book or class will actually teach many of the best techniques, and offer a safe way to develop the elasticity of muscles. With all stretching exercises, always be patient. You may not feel as if you are making any progress, but millimetre by millimetre your muscles will get longer – unless an injury precludes them from doing so. Also remember to keep your breathing going throughout the stretch, as your muscles constantly need oxygen to support them during this type of activity. Find the time to discover stretches for everything from your neck to your calves.

ENDURANCE

Endurance is possibly the most vital quality of your fitness that will be tested during SAS Selection. One exercise should dominate your endurance training – running. Running is a very natural activity, but if you are preparing for SAS Selection you should foster a much more progressive method to building up your endurance than just the odd run of a random distance. Week by week, you need to increase the distances of your runs, the nature of the terrain you run across, and the load you carry. The last element is particularly vital, as there is no point training to do 10-mile runs in training gear with no load-carrying and then be undone when you reach SAS Selection and 36kg (80lb) of gear is hoisted onto your frame.

Like the other forms of training here, be systematic. During each week of training mix the following types of run:

SPEED RUNS

Pick a relatively short course, perhaps 5-8km (about 3-5 miles) and try to complete it at as fast a pace as possible (though take the first few minutes relatively sedately to prevent injury to yourself). Do this initially without any load, but then steadily introduce belt and then pack weights over subsequent weeks. Time yourself carefully and then on the next run try to cover a greater distance in the same time.

Governing your run by time rather than distance means that you do not shorten your training period as you start to improve. This also applies to the subsequent styles of run.

DISTANCE RUN

This is a run of as great a distance as you can manage – it is impossible to give distance recommendations as they will be dependent on your level of fitness. However, you should run for about 90 minutes in total during the first few weeks and then increase this level as your fitness improves. Again, introduce elements of weight as you go along. (The next chapter will illustrate more about how to wear your pack properly for endurance running.)

ENDURANCE RUN OR MARCH

This is where you start to simulate the type of exercises you will face in the SAS Selection training. Pick a varied and challenging route with hills and, if possible, different surfaces (eg grass, sand, gravel). This

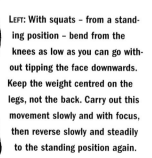

LEFT: With squats – from a standing position – bend from the knees as low as you can go without tipping the face downwards. Keep the weight centred on the legs, not the back. Carry out this movement slowly and with focus, then reverse slowly and steadily to the standing position again.

BELOW: Squats are a good way to develop muscles in the legs, back and shoulders, but make sure that you start off with a weight which you can handle comfortably. Stand straight up with the bar of the weight across the shoulders, and the feet shoulder-width apart.

type of run should generally be carried out with weights and you should allow yourself plenty of time to cover as much distance as possible. This will mean that you have to switch between running and speed-walking to avoid exhaustion. However, when you are walking, it shouldn't feel like a relaxation period – push out long fast steps and keep the arms swinging. Remember what faces you. The 'long drag' exercise towards the end of the SAS 'test week' is a 60km (37-mile) journey by foot over the peaks of the Brecon Beacons which must be completed within 20 hours, with 25kg (55lb) bergen (add to this a rifle and belt kit), through all types of weather, while conducting efficient navigation.

Mix these three types of run each week but remember to leave the rest days for your body to recover. When it comes to your actual running technique you should adhere to the following points about method. Keep as loose as possible in the arms and shoulders (do not clench your fists tightly), and let your body settle into a natural flow and rhythm rather than a forced run. Keep your body in an upright position. Your arms are vital to the running process; swinging them smoothly and in time with your legs will help with both speed and endurance. Make your breathing flow smoothly at all times. Keep your feet quite low to the ground. This reduces the impact suffered by your feet and ankles, which amounts to many tons of pressure over the course of a run.

Running is identified as the most significant endurance exercise for the SAS Selection, but there are several others you should follow. Circuit training is superb for honing endurance and power, as it not only builds up stamina but also strength and flexibility (if the circuit is properly designed). Swimming is also an essential training tool, not least because it is kinder to the body than the impacts of running and circuits. Swimming is part of the SAS Selection course, and you should practise speed swims of 100 metres (100 yards) in various strokes and other more individual exercises such as jumping off high diving boards. Try your hand at swimming wearing a combat uniform and trainers (make sure that someone is with you when you try this).

However you approach your training for SAS Selection, you must have a clear programme in mind. According to former SAS members, too many SAS applicants turn up at Selection and simply expect that because they are generally very fit they will simply rise to the occasion. Some do, especially those that come from units where training is consistently hard, such as the Royal Marines and The Parachute Regiment. But the vast majority do not. Therefore, work out a progressive programme of exercise combining all the elements above which fits into the weeks that remain before your arrival at Hereford. To help you, this chapter contains sample weeks of training from a four-month period of build-up. Use these as a guide rather than as law, because the starting level of fitness varies tremendously between people. But whatever you choose to do, stick to it. The SAS only take people of tremendous self-discipline. So if you cannot stick to your own training course, then you will probably not make it through Selection. Take heed of the words of an NCO from 22 SAS (here taken from Adrian Weale's revealing *The Real SAS*):

'A lot of people think that we are slack somehow, or even sloppy, because we don't all have ultra-short hair and we tend to wear a variety of gear when we're in the field. They get hung up on the sneaky-beaky aspects of what we do and they forget that 22 SAS is just as much a military unit as, I don't know, 3 Para. There isn't a lot of shouting and yelling because there doesn't need to be: almost all the guys are NCOs in their parent units and they know the deal, they know how to behave without a lot of imposed discipline.'
(Adrian Weale, *The Real SAS*, Pan Books, London, 1998)

MENTAL PREPARATION

Before we start to look at the skills you require for SAS Selection, one final word needs to be said on preparation. The mental demands you will face during Selection can be equal to, if not greater than, the physical demands. Thus you should use your period of preparation to get your mind ready as well as your body. Diligently following a self-designed programme of exercise is mental preparation itself, as it encourages self-discipline and an independent attitude, both of which will be required during Selection. Yet there are some more specific tasks which should be followed.

First, accustom yourself to making decisions and solving problems when you are in an exhausted state. This can be done in a variety of ways. One method is to tackle logic problems straight at the end of a long run, and give yourself a tight deadline to solve the problem in order to inject some urgency into the process. In combat, SAS soldiers have to make decisions with a lightning rapidity, and the more you force yourself to make decisions then, like a developing muscle, your decision-making powers will grow.

One useful approach to avoiding decision-making paralysis is to use the US Marine Corps' system of 'rule of three' thinking. When faced by a problem, US Marines are taught to come up with three alternative courses of action – no more, no less – and then choose one and stick with it. While the chosen answer may not be perfect, the result is that positive action is taken and indecision – fatal in combat – is overcome. Practise this in your everyday life until it becomes second nature. Also practise your skills of mental arithmetic – it is not unknown for an exhausted recruit on Selection to reach a Rendezvous point and have the DS give him a mathematical problem to solve under pressure.

Use your preparation period to refine certain skills, such as map reading and navigation. Also use the time to read widely to find out as much as possible about other skills that will be important to you during Selection.

Finally, take time out to socialize. This is vital. Preparing for Selection can become obsessive for some, and they can find that they become rather isolated. This does nothing to help mental agility. Remember that the SAS is also looking for a friendly and cooperative team member, not a reclusive, fitness-obsessed loner. So go out, relax regularly, and approach Selection with an alert and fresh mind.

BELOW: The front crawl works every muscle group in the body. While the legs paddle, stretch each arm forward and bring it down in a scooping motion along the length of the body, eventually lifting out of the water as the other arm reaches forward, taking over the stroke. Twist the head and shoulders in sympathy with the back stroke.

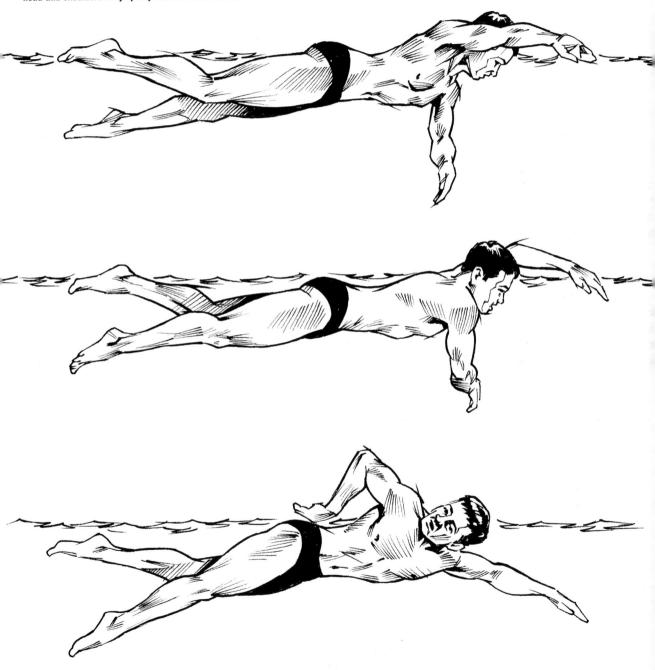

SAMPLE TRAINING SCHEDULES

The following training schedules are taken from Weeks 1, 5, 9 and 14 of a four-month preparation programme. You can use these weeks as landmarks in your preparation programme, and you should work up progressively to each.

WEEK 1

Monday	One hour speed-walk with light bergen load (no more than 5kg/11lb) Gentle 20 minute swim
Tuesday	Weight-training session (light weights) – 30 minutes Racket sports session – 1 hour
Wednesday	Road run of 45 minutes
Thursday	1 hour circuit-training session 20 minutes swim
Friday	90 minutes speed-walk with 10kg/22lb bergen 45 minutes weight training/calisthenics
Saturday	1 hour road run
Sunday	Rest

WEEK 5

Monday	30 minutes stretching session 1 hour hill run (without load)
Tuesday	1 swim 45 minutes Bike ride 45 minutes
Wednesday	2 hour hill walk with bergen, 10kg (22lb)
Thursday	One hour weight training 30 minutes stretching exercises
Friday	Circuit training 1 hour Bike ride 1 hour
Saturday	1 hour road run, working hard
Sunday	Rest

WEEK 9

Monday	45 minutes swimming 1 hour gentle road run
Tuesday	90 minutes run incorporating gentle gradients
Wednesday	Up to 2 hours bergen speed walk with 13kg (30lb) across hills 30 minutes swim
Thursday	1 hour weight training followed by 45 minutes circuits
Friday	90 minutes circuits or hard bike ride up gradients 30 minutes of stretching
Saturday	Up to 2 hours bergen speed walk with 13kg (30lb) across hills
Sunday	Rest

WEEK 14

Monday	Five hours endurance march with bergen (load: 16kg/35lb), ideally over Pen y Fan or other Selection destination. (By now you will be incorporating running into your bergen training) 30 minutes stretching
Tuesday	45 minutes, light circuits 3-4 hours hill training with bergen and substitute rifle
Wednesday	30 minutes gentle swimming 3-4 hours hill training with bergen and substitute rifle at Selection destination such as Black Mountains. Incorporate some navigation practice if possible
Thursday	3-4 hours hill training with bergen and substitute rifle, Brecon Beacons. Load up to 18kg (40lb) 30 minutes stretching
Friday	1 hour gentle cycling or swimming (Around this point you will be building more rest periods into your schedule, as your arrival at Hereford is only a few weeks away and you need to avoid overtraining)
Saturday	4 hours hill training with bergen and substitute rifle, Brecon Beacons. Load up to 18kg (40lb). Keep pace consistently fast and aim to cover about 12 miles
Sunday	Rest

CHAPTER 3

The Skills Required

Being super fit will not by itself be enough to get you into the Special Air Service. Even the first four weeks' training – which focuses very much on the recruits' capacities for endurance – requires specialist mental skills that will be tested at every stage of the recruitment process.

As we start to explore the actual content of the SAS Selection course, we must first issue a note of caution. All military training programmes are in a constant state of evolution, never more so than in the case of elite forces such as the SAS. Special Forces always look to keep themselves at the vanguard of military expertise, and so improvements and adaptations to training methods can occur at varying levels after almost every course.

The caution is essentially this: whatever anybody tells you, or whatever books you read (including this one), do not expect the format to meet your expectations precisely. Caution should even be extended to what former SAS personnel tell you about the Regiment. Contrary to popular perception, due to the explosion of recent SAS publications, all knowledge about SAS practices and training is not in the public domain. Indeed, the commitment to secrecy is such that current SAS members will not readily divulge up-to-date SAS

LEFT: SAS endurance marches are as much tests of navigation as they are of stamina. In winter, even a slight error in direction or location can have fatal consequences. Though navigation is taught on joining the recruitment process, most soldiers arrive with some knowledge.

training methods, even to former members. Approach Selection with an open mind and expect the unexpected. That way you will not be thrown off balance if the course suddenly makes a dramatic deviation from the popularly held formula.

One point we do know about SAS Selection is that there is a distinct difference between the course of the 1950s to 1970s and the programme which is run today. This is partly to do with a culture change in military thinking which occurred some time around the late 1970s and early 1980s. Prior to the 1970s, most commentators of the SAS agree, the generations of menfolk eligible to enter the SAS tended to be more accustomed to hardship than the generations which followed. The reasons are complex, and involve many factors: the dominance of industrial over service industries meant men were more involved in hard physical labour; men of older generations were often brought up outside of the welfare state system and so had fewer material comforts in their upbringing; wartime and post-war austerity tended to condition people towards emotional resilience.

Thus it was that the type of SAS training in the first few decades after the war worked on the basis that men were hardened already, and Selection was

BRECON BEACONS

The Brecon Beacons are a mountainous region of South Wales, and the primary destination for SAS Selection training. Though the mountains are not high compared to other European peaks, they present genuine dangers. The wind speeds are frequently powerful. The weather of the region is fickle. Winter climates in the mountains can have treacherous swings in temperature and precipitation, and even the height of summer can see heavy rainfall and chill night-time temperatures. This unpredictability makes the Brecons a dangerous place for Selection recruits. There have been several fatalities amongst the SAS on the mountains owing to exposure. Many more have been injured by the mountains' variable terrain, which shifts from shingle and bare rock to thick vegetation. Consequently, twisted and broken ankles are commonplace, though more serious injuries have occurred on some of the mountains' steeper gradients.

consequently quite aggressive in the methods used to cull down the applicants to the Regiment. Chief amongst these were the 'sickeners'. Sickeners – as the word implies in its imagery – were exercises introduced early into the Selection course designed to break the spirit and stamina of the recruits. Certain examples are now famous in Special Forces circles. One of the most visceral was getting the recruit to crawl on hands and knees through a trench filled with 0.6m (2ft) deep water mixed with stinking sheep's entrails. This resulted in the recruit having a nauseous stench following him for the rest of the day's training.

Such training methods (as far as we know) are now little practised. Modern generations of recruits to the SAS are not necessarily 'softer' than those of the past (though some veteran SAS drill sergeants might disagree), but on the whole they have had far more comfort in their lives which they have to give up when they arrive at Hereford. Consequently, current SAS Selection tries to encourage the recruits to succeed rather than aim to make them fail. The failure rate is the same as it ever was, and the men who pass the SAS training are as tough as those who forged into the deserts of North Africa or the jungles of Malaya all those decades ago. However, degradation is not a part of the SAS training philosophy today (unless it is for a specific purpose, such as the Interrogation exercises which form part of Continuation Training). The customary greeting that the commanding officer gave new arrivals at Hereford – 'It's nice of you all to come along; I don't suppose many of you will be with us for more than a few days' – has, apparently, also been omitted. All that remains for the recruit is to prove himself through his own self-discipline, toughness and strength of character.

STRUCTURE OF SAS SELECTION

The purpose of this chapter is to outline the specific skills you will need to pass successfully through the SAS Selection course. The next chapter will actually take us through the Route Selection phase of the course, but before we do that it is worthwhile having a straightforward overview of the complete training course which stretches in front of an SAS recruit.

Essentially, the course for joining the SAS separates itself into two halves: Selection Training and Continuation Training. Selection Training is the part of the SAS recruitment process which most soldiers have in their mind's eye upon joining. It lasts about four weeks, and is essentially one of the most intense periods of endurance training of any military unit in the world. The first three weeks are devoted to runs of ever increasing demands in terms of terrain, distance, speed, navigation, load-bearing, and tasking. The level starts high and becomes more difficult – here is where the preparation recommended in Chapter 2 will start paying off.

The Selection process takes place mainly in the Brecon Beacons, a mountainous region of South Wales. Here Selection focuses on navigational skills conducted when exhausted and under pressure. The demands for speed will be punishing, as will the weight of the bergen on your back – 25kg (55lb) of backpack soon saps energy and leg strength when going up one-in-four gradient hill faces for mile after mile. At first you will have group company when you carry out these exercises, but as the course progresses you will increasingly find yourself on your own. This is part of the SAS process of seeing whether you are able to motivate and discipline yourself without an audience or encouragement. By the last week – referred to by some as 'test week' – you will face a quite terrifying series of navigational and endurance tasks across some of the highest mountains in this particular

BELOW: Speed marches vary between running pace and 'double-time', or quick marching. During the early stages of the SAS training the drill instructors want to see people who can keep up with the group – more particularly, those who can stay up with the front of the group.

region. This includes the legendary attrition exercise known as the 'endurance march' or 'long drag'. After this test week, the majority of recruits are heading back to their original units.

To complete Selection training successfully is a moment of euphoria for those few who have managed to stay the distance. Now they can progress on to Continuation Training. This part of the course starts to instruct them in the elite combat skills for which the SAS are famous. Yet caution is needed. Passing Selection does not mean that you are now an SAS soldier. At least 14 weeks of Continuation Training separates you from the famous beige beret and winged badge, and you will be scrutinized at all times. Make a significant mistake, and you will be dropped instantly. The instructors will even make it plain to you that they have failed people on the very last days of Continuation. This is not just scare tactics – it is true.

Continuation Training consists of the following key elements:

- SAS Standard Operating Procedures
- Signalling
- Combat First Aid
- Combat and Survival Training
- Escape and Evasion
- Resistance to Interrogation
- Jungle Training
- Parachute Training

The exact content of these sections will be examined in later chapters. However, a quick glance through the list shows that the demands after Selection training are just as high, though the emphasis shifts in part from physical endurance to mental dexterity. Only at the end of this long process can the soldier put on the SAS beret.

But now we must rewind back to the beginning, and start to look at exactly what skills are required to face the Selection Training and try to make it through to the end.

LEFT: The Parachute Regiment, like the SAS, also conducts endurance training in the Brecon Beacons of Wales. Here a soldier climbs a steep gradient with 23kg (50lb) of bergen backpack pulling on his every step. Any stumble on this terrain can result in a broken ankle.

THE SKILLS REQUIRED

One of the best places to gain a snapshot of the skills required for SAS Selection training is the Special Forces Briefing Course (SFBC) which is run by the SAS as an introduction to the Regiment for prospective recruits. The course is held at Stirling Lines, Hereford, and is open to all those who have been accepted for the full Selection Training. The course begins on a Friday evening and ends about noon on the Sunday. It is highly recommended, as those who attend will have the opportunity to hear from the horse's mouth what the SAS will want from them when they attend Selection proper. You will also get advice on preparation, the format of the Selection course, and receive tips on what to bring. Perhaps most important, you will also get the chance to put questions directly to personnel from Training Squadron and have a look around the SAS base so it is not totally unfamiliar to you when you arrive for proper training. On Saturday night and Sunday morning it is usual for you to have briefings which tell you about life in the Special Forces and whether it is for you.

However, the SFBC is not just a one-way street – you will be evaluated as well. In fact, a poor performance or bad attitude at SFBC will effectively result in you being barred from going on to Selection, so treat the weekend as the beginning of your training and take it very seriously. Apart from the briefing exercises, much of the weekend is taken up with practical tests of your fitness and ability. These tests fall into two basic areas:

FITNESS TESTS
The fitness tests of SFBC do not approach what will be experienced during Selection proper, but they help the SAS instructors to judge whether you have the basic level of strength, stamina and health to attend later training (remember to bring your Fit for Course Certificate signed by your unit medical officer). Most of the fitness tests are fairly standard British Army format Battle Fitness Tests (BFT) and Combat Fitness Tests (CFT). You will be faced with two particular types of runs: a speed run of 2 miles and an endurance-type run of 8 miles. You will be given times to complete on the day, but expect something in the region of 18 minutes and 1 hour

40 minutes respectively for these two runs. Other exercises will involve bleep tests, runs carrying people on your back up and down hills, and common circuit-training routines. The one unexpected element can be the swimming pool exercises. Wearing your standard combat uniform but in trainers, you will jump from a high diving board into the pool, swim 100 metres (100 yards) in 3 minutes and tread water for 10 minutes. This is where your swimming preparation will have paid off.

APTITUDE TEST

Friday evening is usually the time set aside for testing mental suitability for SAS Selection. This will be a series of basic aptitude tests in some of the areas that are key to the SAS – including map reading and

map memorization, first aid, and knowledge of military facts, history and procedures. You will also undergo an IQ test. For this reason it is a good idea to practise solving puzzles in logic for some weeks before you take the course.

The SFBC gives you an idea of the priorities that will face you in Selection, with map reading and endurance sitting somewhere near the top of the list. If you keep your head and work hard, the weekend will do you a lot of good and help place you in the right frame of mind for the training ahead. Just a few words of caution about how you behave during the weekend. You will probably be excited, but keep your nerves under control and do not try to show off – just do what you are told to the best of your ability and leave it up to the DS (Drill Sergeant) to notice you. Be self-confident. Do not follow the others nervously if you think they are doing something wrong, and do not stick anxiously to the heels of the DS waiting for the next order – they are people you do not want to annoy or irritate. As is the case with all SAS training, the SFBC course is designed to see what you are like as a person when placed under pressure, so show 100 per cent commitment throughout.

We will look in more depth at the nature of your behaviour during SAS training in the following chapter, but here we will turn our attention to the skills you will need to take forward to help you have the best chance of passing through Selection.

NAVIGATION

We cover the topic of navigation extensively because it is probably the single most important skill that you will need in SAS Selection training. You may be super fit, with the ability to hump the heaviest of packs up and down the Welsh peaks, but if your navigation skills are poor and you end up at the wrong RV (rendezvous), then you will fail. Furthermore, the time scales are so tight on the SAS endurance marches that if you end up spending too

LEFT: Map reading is a prerequisite for the completion of the SAS endurance phase of training and recruitment. Here a trooper stops to get bearings on a march. His firearm – the 5.56mm M4 carbine – illustrates the SAS's preference for US rifles as standard weapons.

long trying to interpret your map and compass, then you will probably not be able to make up the time on the run. The Welsh mountains are particularly unforgiving. Make a single mistake in your navigation and it could result in you failing Selection, or at the very least cost you valuable reserves of physical energy attempting to put right your mistake. As one SAS NCO put it after completing Selection:

'Navigation is absolutely crucial because if you pick the wrong valley and you end up going up it, you've got to work bloody hard to correct your mistake. You might be going at quite a pace, but it doesn't matter, if you've gone up the wrong valley then you've f***ed up - big style. So your navigation's going to be critical.'
(Adrian Weale, *The Real SAS*, Pan Books, London, 1998)

Do not follow the others nervously if you think they are doing something wrong, and do not stick anxiously to the heels of the DS waiting for the next order

There are two types of navigation with which you should be conversant within the SAS: compass and map navigation and survival navigation. Some books also refer to technology-led navigation such as Global Satellite Positioning systems. These are unlikely to be available to you during Selection so they are not covered in this chapter. For the SAS Selection, the first on this list is by far the most important. Yet here we will also look at survival navigation: firstly, because it can be very useful if disaster strikes during the endurance marches and you lose your compass; and secondly, because it will crop up somewhere in the SAS training programme, if only in Continuation Training.

MAP AND COMPASS NAVIGATION

The starting point for exemplary map and compass navigation begins with a complete appreciation of how to read maps. Maps come in various scales, and the scale you will most typically use on the Selection course is 1:50,000 (this means that one unit of measurement on the map will be equivalent to 50,000 of the same measurements on the ground). You will be issued with good quality Ordnance Survey maps which have an especially detailed presentation of the ground features. Your map is vital to your passing of the Selection course, so treat it with real care. Keep it in a see-through waterproof package to stop it dissolving in the rain. Fold it carefully and always along the original folds, and never mark it. These last two measures are not just to preserve the condition of the map and stop it being obscured by markings. They are also essential SAS security measures – if the trooper is captured by the enemy, there will be nothing on the map which helps the enemy intelligence decipher what the original mission was.

MAP SYMBOLS

When tackling the Selection marches, certain symbols on the map will give the necessary guidance when choosing your route. Contour lines are particularly significant. These are brown or orange lines which join areas of the same height. Contour lines will also give you a clear idea of the gradient of hills and mountains. If the lines are widely and evenly spaced then it means that the gradient is shallow and even. However, the closer the lines are to one another the steeper the gradient – if they are almost touching one another then it means the slope is almost vertical. Before embarking on any journey between two RVs, make sure that you are not blindly selecting an 'as the crow flies' route that will put you into the face of steep slopes which result in lost time as well as the increased danger of injury.

The height of features on Ordnance Survey maps is given by two indications. One is a spot height mark. This is a small black dot with a figure by the side of it indicating height above sea level. The other is a triangular mark which indicates a 'trig' point. A 'trig' point is a physical feature (usually a triangular concrete block) which is used by geologists and cartographers in survey triangulations. On the map the 'trig' point will also indicate altitude.

The next stage to understanding your map is to get to grips with the system of grid referencing. Grid lines, as their name implies, divide the landscape on the map up into a grid of vertical and horizontal lines. Each line is one kilometre apart in real distance. The lines which run vertically are called eastings, and the lines which run horizontally are called northings. Each line is numbered sequentially in the margin of the map, and each square is further divided into 10 increments. This means that any grid reference consists of six numbers, and by using this system you can place your position very accurately on the map. Simply find the point on the map you want to reference, then take the eastings reading first, then the northings reading, and combine the two to make your grid reference. For example, if a mountain summit is in the 24 eastings grid, six tenths along, and in the 57 northings grid, 4 tenths along, then the reference is 246574. (Note: on some OS maps the map is further subdivided into 100,000 metre square divisions and labelled with letters such as SE. This, however, does not change your system of referencing; it simply means that you have to include the letters before the reference – for example, SE246574.) The tenths are often not physically marked, so giving a good, clear reference can rely on your making an accurate imaginary division of each square.

Once you understand your map then for Selection you need to be able to travel quickly and accurately between reference points using your compass. The standard compass for orienteering and navigation is the Silva compass. (Silva is actually the world's largest compass manufacturer – an incredible 1.5 million compasses come out of its factories every year.) They are very tough (they are tested from -40°C to +60°C) and accurate, and made to be used in tandem with detailed map work. There are essentially two navigational procedures you need to be able to perform with your compass: take a bearing (i.e. be able at any point to judge the right direction in which you should be travelling) and also work out your position from two compass readings.

Before we look at these procedures, we need to be aware of the meaning (or meanings) of the word

RIGHT: During the very first weeks of SAS endurance training many of the navigational exercises are conducted in pairs. Here two soldiers work in cooperation, one checking their position on the map while the other soldier takes a directional bearing for the next RV.

'north' in navigation. Confusingly, there are three norths that you may be dealing with. First comes magnetic north, the north to which your compass needle swings owing to the strong magnetic attraction emanating from the north pole. Yet this direction is not directly indicating the centre of the pole itself. That north is called true north. True north is deduced from celestial indicators such as the sun and stars and despite its name, it tends not to be used much for navigational purposes. Finally comes grid north. This is the north indicated by the grid lines on your map, and is the north from which we take map readings.

Into this equation we also have to compute the fact that magnetic north itself is not a fixed entity. The magnetic field of the earth shifts according to where you are, the year, and what time of the year. To gain an accurate reading from your map, you will have to calculate the magnetic variation and then add or subtract this from your directional bearing. A compass is marked out in 360 degrees and/or 6400 mils. Mils are the usual measurement for calculating magnetic variation. Thankfully, OS maps contain the details of magnetic variation and you can use this to subtract or add the right figure to gain an accurate map reading.

If your map and compass handling is slow, then each reading taken in the field will chip valuable minutes off the time available for you to reach your next RV

As stated above, the two central map and compass activities of Selection are taking a bearing and finding your position. To do the former, first place your compass flat on the map, with one edge of the compass (Silva compasses are rectangular) running from point A of your journey to point B. Next, turn the adjustable dial on the top of the compass until the north-south lines are parallel with the meridian lines of the map and the north arrow points to map north. At this point you should then add or subtract your magnetic variation. Next, take your compass off the map. Do not touch the dial settings you have just set but instead turn the entire compass horizontally until the magnetic needle itself points to the north indicator on the dial and is also parallel with the compass's north-south lines. The directional arrow on the compass will now be pointing in the direction in which you should travel. Once you have your bearing then it is simply a matter of regularly consulting your compass to make sure that you are keeping on the right track all the time.

Of course, the accuracy of this method for navigation depends to a large extent on knowing where you are in the first place or along your route. This

LEFT: The latter phases of endurance testing involve overnight marches and camping out. Note the sensible site these troopers have chosen, with the wall to their backs to avoid the potentially lethal wind chill blowing over the summit of the hill.

of your compass towards a feature which you can clearly identify on the map. Then turn the compass dial until the north-south lines run parallel to the needle and the red needle tip is pointing in the same direction as the north marker on the dial. Then add or subtract the magnetic variation.

Next, take your map and place the edge of the compass so that it touches the symbol for the feature you have just taken a bearing from. Then turn the whole compass, pivoting on this point, horizontally until the north-south lines are parallel to the north lines on the map and the magnetic needle is pointing to north on the map itself. Next, draw a line along the edge of your compass. To complete the process, find another identifiable feature on the landscape in front of you and repeat the whole process again. Where the two lines bisect each other indicates your position.

Before attending Selection, practise these important techniques of map and compass handling until you feel it has become second nature to you. If your map and compass handling is slow, then each reading taken in the field will chip valuable minutes off the time available for you to reach your next RV. One other thing – remember to take a spare compass with you. It is not uncommon for soldiers to lose their compasses during the confusion and scrambling of Selection. Having a simple button compass tucked away in a pocket can be a Selection saver in these instances.

SURVIVAL NAVIGATION

Map and compass navigation is one of the most precise ways to find your way around the Selection course. However, it is advisable that you also learn some basic techniques of survival navigation. Whether in summer or winter, Selection will run whatever the weather and conditions in the Welsh mountains, which can be harsh and unforgiving. Even some experienced SAS men have become fatalities during Selection (see the next chapter), so losing your way can have dire consequences.

can be particularly tough during SAS Selection. You will be physically exhausted and the weather in the Brecon Beacons is notoriously fickle. Visibility can suddenly be reduced and that's when confusion and mistakes can take hold. Thus an essential compass skill is locating your position by a technique called resection.

Resection consists of taking two compass readings from identifiable features and using these readings to produce an exact plot of your position on the map. First, point the direction or sighting arrow

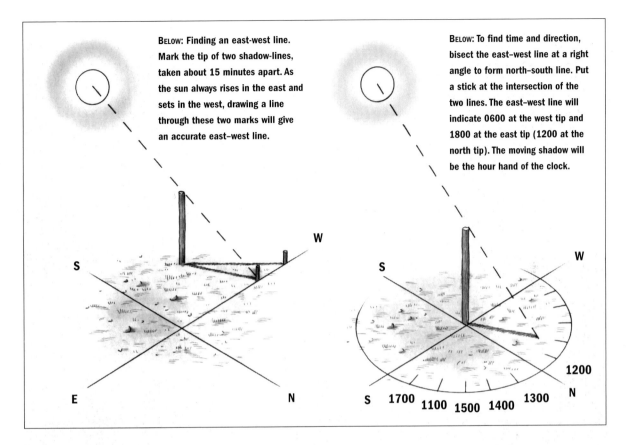

BELOW: Finding an east-west line. Mark the tip of two shadow-lines, taken about 15 minutes apart. As the sun always rises in the east and sets in the west, drawing a line through these two marks will give an accurate east–west line.

BELOW: To find time and direction, bisect the east–west line at a right angle to form north–south line. Put a stick at the intersection of the two lines. The east–west line will indicate 0600 at the west tip and 1800 at the east tip (1200 at the north tip). The moving shadow will be the hour hand of the clock.

As previously mentioned, it is possible to mislay compasses (even maps) during Selection, so having a few of the essential points of survival navigation might help you to get to safety, if not the next RV. But survival navigation has a more general benefit in that it can help you between map and compass consultations, giving you directional guidance at a glance and thus speeding up your travel times.

Two of the most readily accessible tools to survival navigation are trees and plant life. In exposed regions such as the Brecon Beacons, the growth patterns of plant life are distinctly shaped by the elements, and they provide us with some useful indicators of direction. Before you attend the course, find out from someone very familiar with

LEFT: Regular checking of the map is essential to avoid any costly mistakes during endurance marches. Ideally, the map and compass should be consulted every 10 minutes, to confirm the direction of travel and checking that obvious landmarks are where they should be.

the area (for example, from a mountain-rescue organization or hill-walking club) the direction of the prevailing winds. Isolated trees or trees at high elevations will be shaped by these winds – if the prevailing winds blow from the south-west then the tree will be bent over to the north-east and so on. (Be very careful that you do not misread the signs – a tree may be located in a peculiar position in which the prevailing winds are essentially being re-routed by a valley or hillside.) Reeds and flowers also tend to have their flower growth on the leeward (sheltered side) of the plant. Mosses generally grow on the north side of rocks because they like to avoid heat and sunlight and thrive on darkness and moisture.

Other useful navigational signs from plant life derive from the fact that in the UK the sun always tracks across the south (this in itself can be a good indicator that you are travelling in the right direction – if you are meant to be heading north but the

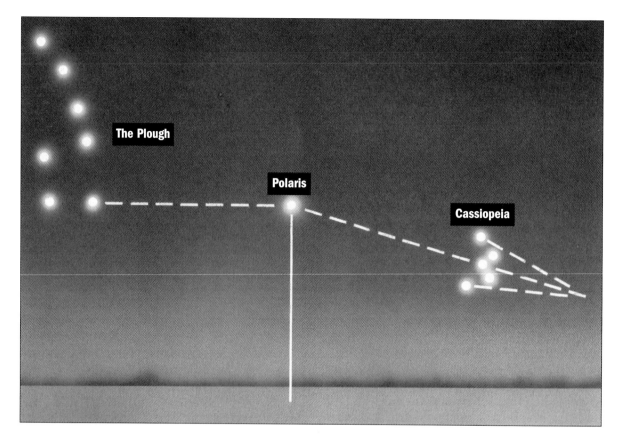

The Plough

Polaris

Cassiopeia

ABOVE: Polaris, or the North Star, will give the observer an accurate celestial indicator of a northern direction. Follow through the last two stars at the crook of The Plough, or alternatively through the centre star of Cassiopeia, as shown, in order to locate it.

sun is in your face then you have obviously made a mistake navigating). As for plants, the fact that branches and leaves stretch out towards the sun can be useful. Horizontal growth of branches from one side of exposed trees often indicates south (to check this, observe the uppermost branches; they will often seem to stretch over towards the sun).

Other techniques can be applied to the sun to make more specific measures of east-west and north-south lines. The two most common techniques to note are the shadow-and-stick method and the watch method. The former is very simple, but it does rely on clear, consistent sunshine to be effective (something far from guaranteed in this part of Wales). Place a stick about 1m (3ft) high into the ground, and mark the tip of the shadow with a stone. Then wait for about 10 minutes until the sun shifts in the sky. Mark the new position of the tip of the shadow with another stone. Drawing a line between the two stones will give you an east-west axis, and bisecting this at a right angle will produce a north-south axis. Thus on the ground you will have produced the basic compass points which can give reasonable guidance as to direction.

The watch method is better, mainly because it is quicker (there are few occasions on Selection when you can afford to wait around for 10 minutes). Also the mechanism for gaining the direction is on your wrist – unless you are wearing a digital watch, which is not suitable. Point the hour hand at the sun. True south is the point halfway between the hour hand in this direction and the 12 numeral of the watch. (Note: this is the watch technique for the northern hemisphere only. In the southern hemisphere the 12 numeral itself is pointed at the sun and true north is indicated by the mid-point between the 12 and the hour hand).

The last two methods of survival navigation obviously depend on a crucial factor: that it is daytime. As Selection will take you through the night in many instances, other methods must then come into play. Hence you need to swap the sun for other stars. Star identification can be extremely useful during Selection, not least because being up in the mountains means there is no light pollution and the stars can be highly visible – again depending on the weather. In the northern hemisphere, your most valuable guide is undoubtedly the Pole Star. This can be tracked from the constellations The Plough and Cassiopeia. The Pole Star will give you an unerring indication of north, wherever you are in the northern hemisphere, and so can act as a useful fixed point of guidance throughout the night phases of the Selection process.

One further method of using the stars for navigation involves watching star movement. For this you will need to find a position in which you can sit relatively still. From this position, pick two fixed points on the horizon or above you. Looking at these, note the position of stars in relation to the points over the next ten minutes or so. The movement of the star in relation to the points will give you a general indicator of the direction in which you are facing:

Star rising – east
Star falling – west
Star loops towards right – south
Star loops towards left – north

Obviously, all these methods of survival navigation are far from infallible and should be used with caution. Yet remembering them can help you to speed up the whole process of navigation and keep you moving towards your next important RV.

KIT – CONTENTS AND HANDLING

The SAS Selection course would be bad enough if it were conducted in light PT gear and training shoes, but add the weight of full combat uniform, combat boots, a dead-weight bergen, and a rifle, and you have a recipe for torment. Handling and choosing

MOON NAVIGATION

Though the stars are probably the most accurate celestial body to give you night-time navigational help, the moon may also give you some handy directional indications. Moreover, the greater luminescence of the moon means that it can even be seen through moderate cloud cover, whereas stars under those conditions will be totally obscured.

The moon's directional guidance is very general. However, its cyclical orbit produces different patterns of illumination on its surface as it turns in relation to the sun. It is both the pattern of illumination and the time of its visibility that gives us a navigational reference. In the tail-end of daylight, if the moon rises before the sun has actually set, then the side which is illuminated will be facing west. After midnight, the illuminated side will be facing east. Quarter moons also give some navigational help. In your mind's eye, draw an imaginary line through the horns of the quarter moon so that it touches the horizon. In the northern hemisphere the place where it touches the horizon is due north. It is due south if you are in the southern hemisphere.

this kit properly is vital to aid the success of your attempt at Selection. One SAS soldier recounts an applicant turning up for Selection in brand new boots and a pair of nylon socks. The resulting blisters were so horrendous that they soon put paid to his chances of passing. Such an example may seem obvious, but think hard in advance about what you take on Selection as it will pay off when training starts. (As with all recommendations in this book, check with the SAS first about what kit is appropriate before you arrive at Hereford, as specifications and procedures may change from year to year.)

CLOTHING

As we have just given an example of poorly selected boots, a good place to start is footwear. Given the example, it is unwise to arrive with a pair of boots which are not broken in. Take two pairs of good quality, heeled combat boots, plus two pairs of training shoes for additional exercises around Hereford. Your boots are the single most important part of your kit, and you will have long hours of marching ahead of you, so make sure that you look after them. Apply a good waterproofing, rubbed well into the leather over several sessions. In particular, make sure that you dry them out properly after they have been soaked (which they often will be during Selection). Some of the worst effects of soaking can be prevented by wearing waterproof gaiters, either full-boot or uppers-only variants. The full-boot variant is the best, as this can even stop water from penetrating the boot when crossing streams. However, at some point your boots will be soaked. To dry them out put them out in the sun or near an artificial heat source and allow them to dry slowly (stuffing them full of newspaper can also help draw the wet out). Do not place them under

Most of your body heat is lost through the extremities – head, hands and feet. Keep these vital areas as warm as possible in the wintertime

strong heat – this will most likely result in the leather cracking open. Also remember to dry out the laces on their own (make several pairs of spare laces an essential part of your kit for Selection). Take several pairs of good quality woollen socks and change them regularly, especially if you feel blisters forming.

Moving on to the uniform as a whole, you will in a sense have little choice about this as the SAS use the regular DPM (Disruptive Pattern Material) British Army uniforms, so your uniform will be like that of any other soldier. However, you should pay special attention to the make up of your clothing. Temperature control is vital on the long and exhausting marches up and down the Welsh mountains, exposed to everything the seasonal elements can throw at you. If you become excessively hot, you could fail the course (or worse) through heat exhaustion. Become too cold, and you might even start to suffer from hypothermia. (These two problems can even occur during the same exercise – for example, if you get too hot in the winter and sweat profusely, when you stop the sweat can freeze in

your clothing and precipitate cold-related injuries and illnesses.)

The key thing to remember is to wear plenty of layers so you can control your temperature with some precision. Many thin layers are warmer than few thick layers, and have the virtue of you being able to shed or put on items of clothing as your body tells you (more about coping with medical problems will be covered in the next chapter). Start with a base layer like a T-shirt. Some recommend cotton, yet cotton T-shirts act like sponges for sweat and they increase the risk of freezing in cold climates. If you can, invest in some of the modern civilian mountain wear which allows sweat to 'wick' away.

Next, follow up with a good woollen sweater, and over the top of these will go your DPM jacket and waterproof if need be. Whatever the clothing combination, remember several vital points. Most of your body heat is lost through the extremities – head, hands, and feet. Keep these vital areas as warm as possible in the wintertime. Surprisingly, a great deal of heat can be lost via open, flapping cuffs and waistbands of jackets, as well as at the ankles of trousers. If at all possible see if you can get hold of clothing with fasteners or elastic at these points, or even slip a set of elastic bands into your kit in case you need to improvise. As with your boots, make sure that your clothing is well maintained. Repair it as soon as it becomes damaged, and above all keep it as clean as possible – dirty clothing is not as efficient as clean clothing in trapping warm air. Hang wet clothing from branches or over dry, warm rocks to air dry.

BERGEN

The bergen is of course the workhorse pack of the British Army. The bergen backpack used to be a very uncomfortable monster indeed, and from its introduction to the SAS during the Malayan campaign, generations of SAS troopers were forced to struggle with its painful load-bearing properties. Eventually, the source of most of the discomfort – a rigid metal

RIGHT: The clothing issued to you for SAS training will give you good protection throughout the course. Goretex outer clothes ensure that perspiration evaporates while rainwater is not let in. Store protective clothes at the top of the bergen for immediate use when required.

LEFT AND ABOVE: Essential waterproofs and tent rolls are kept at the top of the bergen where most accessible. If clothes do become wet through, try to dry them out at every major stop. Change your socks whenever the feet feel chilled, or regularly if they have severe blisters.

'A' frame – was removed in the early 1980s and the softer Cyclops and Crusader bergens were introduced to the SAS.

Reading any tale of SAS survival almost always reveals the total dependency on the bergen. Standard belt kit contains survival gear, but the bergen has everything the soldier needs to survive in the field with a relative degree of comfort. However, the plethora of both combat and survival equipment carried by a Special Forces soldier means that the bergen contents often add up to an appalling weight. It is your ability to carry this weight – anything around 40kg (88lb) is possible – which is tested during Selection, and also your ability to handle your equipment sensibly during the endurance tests.

A primary skill in bergen handling is how to pack and unpack. The worst approach you can take is just to throw everything you need into the bag without a structured idea of where items are going. This can result in lost time, damaged equipment and clothing, and frustration you could do without. The first rule is to pack all the items you will need regularly – such as food packages and cooking equipment – in the bergen's side pockets for easy access. On the other hand, items which you will only need once or twice during the day – such as the sleeping bag – will go towards the bottom of the pack. Use the top flap of the bergen to store wet-weather clothing so it is quickly accessible in case of a downpour.

The bergen is made of waterproof material, but this does not mean that your contents are totally protected from a heavy downpour. Pack the internal contents of your bergen in several separate strong plastic bags so that all items are protected from a sudden soaking. This is especially important

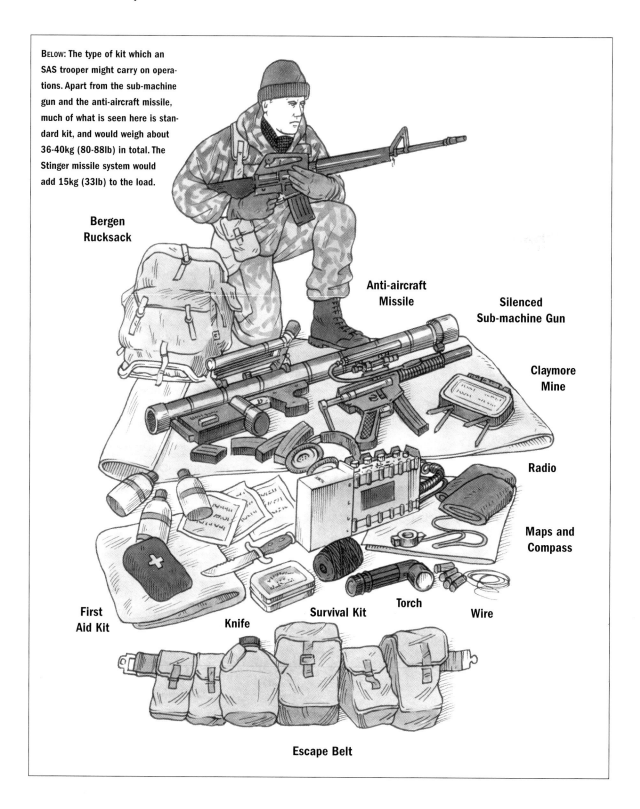

BELOW: The type of kit which an SAS trooper might carry on operations. Apart from the sub-machine gun and the anti-aircraft missile, much of what is seen here is standard kit, and would weigh about 36-40kg (80-88lb) in total. The Stinger missile system would add 15kg (33lb) to the load.

Bergen Rucksack

Anti-aircraft Missile

Silenced Sub-machine Gun

Claymore Mine

Radio

Maps and Compass

First Aid Kit

Knife

Survival Kit

Torch

Wire

Escape Belt

If you treat your bergen well and pack it using your intelligence, this will make your rucksack more of a friend than an enemy

for items such as sleeping bags which may be ruined if they become completely saturated.

After you have finally packed your bergen, you then have to wear it. Take time in the early days of the Selection course to get used to the feel of your bergen and keep adjusting it until you achieve maximum comfort. Watch out for points which are rubbing (these can become very blistered during the endurance marches) and sensations of numbness in limbs (straps may be too tight). Adjust and pad your bergen until it fits you perfectly.

Regarding the right way in which to carry your bergen, the best position is with the bergen high up on the back and the weight distribution slightly heavier on the shoulders than on the hips. Make sure it is also central. By keeping your bergen higher it means that your legs take more of the strain, thus reducing some of the risk of back injury from leaning against the straps too heavily for too long. In addition, watch your running technique – use long, loping steps when running with the bergen as this stops the painful 'bergen bounce' upon your back for mile after mile.

If you treat your bergen well and pack it using your intelligence, this will make your rucksack more of a friend than an enemy. However, there is no way around the fact that the bergen will become torture itself after you have marched something like 20 miles up and down tough gradients into hard-blowing winds.

ADDITIONAL KIT ITEMS

Although the bergen rucksack used to be filled with bricks in the old days to ensure a standardized weight for everyone who was undergoing Selection, today proper equipment is used to gain a more faithful insight into the candidates' ability to carry combat loads. Much of the kit which you take with you on SAS Selection will be the standard kit of a British Army infantryman – everything from ammunition to cooking stoves. However, because

the demands of SAS Selection are so unique, there are one or two additional items which you should take with you that will assist you in coping with the course.

Several of these items have been mentioned already. A button compass is an almost weight-free supplement to your regular navigational tools, but it is one for which you will be immensely grateful should you lose your regular compass. In this regard taking a magnet and pin can also be a good idea. If you stroke a magnet down the pin in one direction only, it becomes, in effect, a compass needle. The best way to get a reading from this is to push the magnetized pin through something which floats – such as a cork or a few matchsticks – and float it in water in a non-metallic container. The pin will need regular re-magnetizing, but it is an ad hoc method which may save the day if you lose all other compass tools or instruments.

Navigation can also be assisted by taking a luminous light with you, such as those used by mountain climbers. These are readily available from climbing and outdoor pursuits suppliers. Because they produce their light through chemical process rather than battery power you can be ensured of a light source powerful enough to read your map and compass at night (some tie it around their wrists so that the light source is always present, but experienced Special Forces soldiers would generally want to avoid such a highly visible signature). Remember to take a good, waterproof map case with you, as protecting your maps during Selection will of course be your own responsibility.

Next come bergen-related items. These include foam padding to put under painful straps, and also plastic bags (ideally bin bags without any moulding holes in them) to keep items of kit waterproof. Remember to take with you materials you can use to carry out temporary repairs to your kit – particularly strong masking tape and a tube of superglue. Other handy items include strong elastic bands and the ever-useful Swiss-Army knife.

In this chapter we have looked at how to handle navigation, clothing and equipment-carrying. This will stand you in good stead when you launch into the Selection course. The content of the SAS Selection course is the focus of the next chapter.

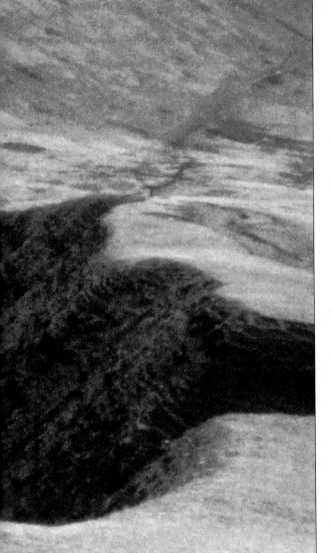

Selection

Selection is a four-week period which in itself would test the stamina of a professional athlete. The intention behind this severity of training is partly to see whether the recruit can combine physical toughness with mental tenacity, even when operating alone and far from colleagues.

This chapter on the actual SAS Selection course begins with a cautionary tale to show why it should command your utmost respect. It is worth remembering that the type of exercises you will perform on Selection are not solely designed to test your powers of stamina and endurance. Rather, in many ways Selection is aimed at revealing your capacity to survive. Even the most experienced and strongest of SAS troopers still approach the Selection routes with some degree of trepidation, and to see why we need to look at the final acts in the life of Major Mike Kealy.

Mike Kealy was, and remains, a legend within the SAS hall of fame. He rose to prominence following his leadership of the heroic defence during the Battle of Mirbat in Oman in 1972. On 19 July of that year, over 250 Adoo (enemy) guerrillas launched an all-out attack against the port of Mirbat, Dhofar province, southern Oman. Kealy, then a Captain, led a squad of nine SAS troopers in a truly epic action which resulted in the attack being repulsed and 38 Adoo guerrillas being killed. For his part in the battle Kealy was awarded the Distinguished Service Order.

LEFT: A sobering view of the Brecon Beacons. There are many harsher mountain environments in Europe, but the Brecon Beacons are extremely variable in their climate – winter weather can turn from temperate sunny days to sub-zero sleet and snow in a matter of hours.

Seven years were to pass before Kealy, then a Major, decided that he needed to test himself once again. At this point he was nearly 33, yet despite his age he decided that he would attempt once again the infamous 'endurance march' of the SAS Selection course. This march, as we shall see later, is the apogee of Selection, a truly frightening slog over the Welsh mountains with everything set against you – weather, kit weight, time, and gradient. Kealy's desire was to prove that he still had the resources to complete this most sapping of exercises according to the time-scale he had to meet when he took Selection all those years ago. Up to this point in SAS history, the Selection course had claimed the lives

of just three volunteers – not a bad rate when considering that Selection had been running for 23 years by 1979. Yet although Kealy was no mere recruit, he was about to join the list of fatalities.

Major Kealy set off from the Talybont reservoir on 1 February 1979. The weather – Kealy having chosen to go for a winter march – was appalling. Temperatures kept dipping below zero, rain swept the landscape at right angles before turning to snow, sleet and ice. Visibility was appalling – a crucial factor which made accurate navigation that much more significant.

Kealy was actually part of a group of 30 recruits, separated into two units of 15. However, Kealy

him, Kealy was in danger from the first moment he strode out from the start point at a fast pace.

With each increment of altitude gained, Kealy's situation went from bad to worse. At first he had overtaken all the others starting out, but as the hours ticked by more and more of the recruits started to pass an increasingly exhausted-looking Kealy. The weather was abominable, and all present on the course that day – apart from Kealy – realized their lives were in potential danger from hypothermia. After consultation amongst themselves, a large body of the recruits decided that it would be advisable to go and shelter in a barn lower down in one of the valleys, out of the freezing, biting wind. Kealy chose not to go, and the group saw Kealy swallowed up by the sub-zero cloudbase.

Eventually, rear elements of the training party came across a shattered and hypothermic Kealy. Because he had no waterproofs, the freezing rain and sleet had soaked him through, the powerful wind-chill then sucking the heat from his body. The soldiers attempted to help Kealy, but Kealy was strongly resistant to their attentions. Though the soldiers attempted to draw Kealy off to safety, night had fallen by this time and Kealy was once again lost. The two recruits who had tended to Kealy spent an hour in search of him, but they too were eventually driven down to seek shelter in the barn.

At around 1000 hours the next morning, two SAS troopers on the march came across Kealy. He was barely alive, sat on the ground and so covered in snow that at first the soldiers thought he was a snow-blanketed rock. While one soldier tended to the now dying man, the other set off to raise help. Such was the atrocious nature of the weather involved, that Kealy's location was not discovered until 0430 the next day. After another night on the mountain – despite a brave SAS corporal staying in a snow-hole with the dying officer – Kealy's condition was hopeless and he was already dead when the helicopter airlifted him from the mountain. In retrospect, we can see a catalogue of errors on

chose to stride out alone. In retrospect, this was a decision which probably killed him. It was not the only poor decision Kealy made that day. His enthusiasm overrode his judgement. He not only omitted combat waterproofs from his pack, but also did not take simple precautions such as wearing the gloves in his pack. The pack was a problem in itself – Kealy had made up the required weight of 23kg (50lb) using bricks. Though this used to be common practice in the early days of the SAS, it was eventually abandoned because bricks form an uneven weight distribution in the pack, which can destabilize the carrier and give an unrealistic test of their load-carrying abilities. With all these factors stacked against

Kealy's part which directly contributed to his death, though we should also remember that other members of the group were also treated for exposure, even some of those who had sought shelter. Yet Kealy's fatal attempt to accomplish the march (he had spent the immediate preceding years in administrative work), is part of the SAS culture of absolute self-reliance and toughness. The premise is that training should be as tough as the reality of actual operations, and on operations the SAS soldier has to stand on his own two feet or die.

Yet the tragic demise of Major Mike Kealy has an important lesson for us regarding Selection. Selection has to be approached with the utmost seriousness, and it requires judgement as well as physical fitness to complete successfully. The four-week period of Selection could make genuine demands upon your ability to survive, and so your preparation should take on an added diligence. You should also listen intently to all advice from experienced NCOs and officers. Kealy was in effect killed by the atrocious winter weather, but in whatever season you attend the Selection course there are dangers ever present, as this SAS NCO points out:

'Summer Selection has a few advantages over and above the winter months: that the weather is generally more congenial; you don't get so many sleet storms coming straight into your face which is very depressing; you get a generally more pleasant run. But the downside is that the surface isn't easy to cover up in the hills because all the growth is up: all the plants are up, the grass is much longer and the 'babies' heads' are in full flow – one mate of mine actually dug up a baby's head and brought it back with him 'cause they're such nasty little b***ers – they're little wavy tufts which ensure that you destroy your ankles with consummate ease.'

A wrecked ankle in the summer months can be just as dangerous as a wrecked ankle in the winter in some climates, though in the Brecon Beacons it is more likely to result simply in your failing Selection. However, the point is that a recruit about to undergo Selection will have to face the realities of a hostile environment. Selection is consequently far from predictable, and whatever recommendations follow, you should always think for yourself and make your own judgements as you go along.

JOINING SELECTION

The Selection process usually begins at the Stirling Lines base in Hereford. On your first day do not be late under any circumstances – if you have to travel far it is probably better to go to Hereford the night before and stay in one of the plentiful B&Bs in the area. The next morning you can get a taxi to the base (ask the driver to take you to Main Gate, Stirling Lines) refreshed and ready. Once at Hereford you will be given a briefing about what is going to happen, and also a 'welcome' from the commanding officer. Don't be surprised if the tone

The four-week period of Selection could make genuine demands upon your ability to survive, and so your preparation should take on an added diligence

of the officers and NCOs is distinctly cold towards you. At this stage all they see in front of them is 150 unproven young men, the vast majority of whom will have disappeared from training by the end of the four weeks. They have little commitment to you at this stage and they can send you back to your unit at any time during Selection and Continuation Training.

The first day or so may be spent in administrative processing, as well as performing various qualifying fitness tests such as road runs and PT exercises. Your first physical test will come in the form of the basic Army Combat Fitness Tests (see previous chapter). These you will have encountered already in your home units, but make very sure that you perform them to an exemplary standard when under the watchful eyes of the instructors. Stay up with the lead groups at all times. The CFT in itself will serve to weed out a few individuals who have not prepared for Selection properly. Yet this stage of Selection is just the warm-up to the main part of the course, over three weeks spent in the mountainous areas of Wales, particularly the Brecon Beacons, on endurance and navigation exercises. This is Selection proper, and will culminate in the fearsome 'long drag' or 'endurance march' which proved to be the undoing of Kealy.

The base destination for this phase of Selection is usually the Sennybridge camp in Powys. Sennybridge is not only the central location for Selection, it is also one of the SAS's most active live-firing ranges, so expect plenty of activity there from recruits and serving troopers alike. Sennybridge serves an important purpose for Selection, in that it stops the regular travelling from Hereford which SAS recruits had to undertake before Sennybridge became the Selection HQ in the late 1980s.

Treat the maps like gold. They will be your guidance across the routes of the next weeks, and mistreating them could have serious implications

Once you arrive at Sennybridge, you will receive the various items of kit that will be essential over the coming days. As for clothing and load-carrying equipment, your standard combat uniform will be supplemented by waterproof jacket and trousers, and also your trusty bergen. The bergen will come with a luminous panel which acts as a visibility marker in case you have to be located and rescued during Selection. Survival is very much the emphasis in the kit you receive at Sennybridge. As well as packs of emergency rations you will also have rescue flares, a homing beacon device to attach to your bergen, a torch, a sleeping bag and 'bivi' (bivouac) bag, plus a survival suit. At this point you will also receive your rifle. As with all kit, check it over thoroughly to ensure no items are damaged. Fit your bergen diligently, checking for any parts that are rubbing and adjust or pad out accordingly.

Part of your kit issue at Sennybridge will include a Silva compass and also five 1:50,000 Ordnance Survey maps of the areas you will face during Selection. It would be a good idea to get hold of these maps before you actually arrive at Selection. The particular maps you will need are of the Brecon Beacons, Black Mountains, Hay on Wye, Elan Valley, and the Forest of Dean. Study the ground that you will face and note its features, especially the contours and gradients and useful navigational features, such as valleys, streams and rock outcrops. Part of the skill of map reading is to be able to look at a map and 'see' the 3-D landscape in your mind's eye. Treat the maps like gold. They will be your guidance across the routes of the next weeks, and mistreating them could have serious implications – it would be tragic to fail simply because you folded the map the wrong way, split a fold, and thus obscured an important feature. Because the SAS come from all sectors of the British armed forces, not everyone arrives at Sennybridge with an equal understanding of navigation. Thus the instructors at Sennybridge will run through navigational essentials. Listen hard, as they understand the peculiar direction-finding problems that you will face in the Brecon Beacons and their guidance will be invaluable.

Once you have been issued with your kit and you have been briefed on what is ahead of you, then you must start the Selection routes.

SELECTION ROUTES

Unfortunately, any book which aims to teach about SAS selection has to concede that there is no way of knowing the routes that will be taken in any particular year. Every year the routes are changed and adapted, and former soldiers are not willing to reveal the exact routes taken during their Selection. This is as it should be, for if the routes were known then the security of the SAS would be severely compromised. However, what we can say is that the nature of the Selection routes is fairly predictable.

Most Selection courses begin with speed marches of 16-24km (10-15 miles) with something in the region of 18kg (40lb) of pack plus belt and rifle. You will go out in groups accompanied by a DS (Drill Sergeant), and be given a route with about two or three rendezvous points. The governing factors in the route selection at this stage – and indeed throughout Selection – are speed and accurate navigation. Keep up with the DS if possible. This will

not be easy as he will be extremely fit and also be carrying only a pack, but keeping up with the lead group will buy you if not his approval then at least his resistance to ejecting you from the course. Most of the courses at this stage will involve steep gradients, and they will only get worse and longer as the course gets on.

One of the most popular destinations during Selection is the mountain called Pen y Fan (pronounced 'Pen-a-van'). At 886m (2900ft), Pen y Fan is the highest peak in the Brecon Beacons and is capped by a long, stony and exposed plateau. The ascent up here is arduous to say the least, but at some point during SAS Selection you will have to tackle the infamous 'Fan dance' – basically this exercise involves running up and down Pen y Fan's bleak slopes fully loaded with kit. Here's what one SAS trooper who successfully completed Selection had to say about this element:

Above: Whatever you do during Selection, do not skimp on food. If time is short, eat on the move, but make sure that your calorie intake is kept high through frequent small meals, otherwise concentration and energy will soon slip away and the danger of exposure becomes real.

'The real test is "Fan dance" which is a beast – really hard work. No matter how fit you are it's hard work, and anybody who says it isn't is full of s**t. In theory you've got to follow your DS, and if you stay with him you're guaranteed to be one of the fittest blokes there because the instructors are very fit and they are not carrying very much kit: they're not carrying a rifle, just a pack. It's basically a run up Pen y Fan: up the hill, down the other side, back up Fan and back down it again, and at the end of that your legs feel like b***dy jelly.'

As this soldier later goes on to recount, the 'Fan dance' really starts to cull the numbers of recruits.

Remember one thing about the Selection process – you drop out and take too much of a rest, you miss a rendezvous point or time, you show that you cannot control your emotions, then you will be dropped from the course instantly and RTU'd (returned to unit). SAS Selection has to be done right all the time, every hour, if it is to be passed. This is why each Selection course yields few more than 10 successful recruits:

'By the time we'd formed up again after the Fan and had another briefing before we went on the hills again, we were probably down to half the course, so we'd got through half the people and this was day two of the course. Half the people simply didn't make the grade on the 'Fan dance'. Logistically they have to do it - they simply can't feed and transport that many people around - so there is a practical reason, but also, frankly, if they can't pass the Fan then they won't be able to do the rest of it. The DS's are brutal about it, they can set the times how they like and they don't have to give any reason - it's just "F**k off, you failed!"'

As we can now see, your first few days of route Selection are make or break, so you must throw yourself into the exercises with all your energy and total commitment.

The first week will consist of similar exercises, always in groups and led by a DS. Deviations from this routine during the first week will consist of your day out for map-reading tuition (though the afternoon will probably involve some fairly horrendous PT exercises in a gym) and also a trip to the swimming baths for water-based training. This latter part of training can involve having to swim 20 lengths and 20 widths in uniform, boots, belt and water bottle, and then performing other tasks such as underwater swimming with the same kit and treading water exercises.

By the end of the first week you will have noticed that you are having to perform the navigational and

LEFT: By the third week of Selection, the recruit will find himself increasingly alone during the endurance marches. The instructors are looking for a person who is not intimidated by long periods of isolated decision-making and lack of social contact.

endurance exercises in smaller and smaller groups and without the leadership of the DS. Indeed, some of the final exercises of the first week are most likely to be conducted in pairs. You will also notice that the number of rendezvous points will be increasing steadily and the times to reach each decreasing steadily. This is the direction in which matters will progress over the weeks to come. Indeed, however tough the first week has seemed it will get much, much worse.

ENDURANCE PHASE
During the second and subsequent weeks of Selection training, the endurance element of the routes will reach almost unbearable levels, culminating in a final dramatic period called 'test week', a period of seven days which features the summit of Selection - the 'long drag' or 'endurance march'. Self-reliance is the quality that the DS is looking for during this phase. You will have to navigate without fault and more and more of the exercises will be conducted on your own to see if you have the character to survive long periods of isolation (a quality which you will need to possess for SAS operations).

The second phase of the Selection process will also test your levels of endurance to their limits. Any minor injuries picked up over this period are likely to become searing problems in their own right, particularly any blisters, bruised heels or twisted joints. If they are genuinely disabling, then the DS will be looking for you to summon up the ability to ignore the pain and push yourself through to your very limits of endurance. If they cannot be tolerated - either for psychological or medical reasons - then you will be RTU'd without a murmur. Only the toughest will survive.

Regarding rest days, during Selection in its entirety you are likely to receive about three full days off, though you will remain on camp and will probably have to attend briefings on those days. You will probably also receive a one-to-one talk with a DS who will give you a comment on your performance and tell you where you need to improve and what you need to do to keep on course. Make sure that you get plenty of rest on these days, however. They are usually placed just before periods of extremely challenging exercise and they are designed to help you regain some of your energy. Watch how you socialize on these days especially. By all means go out and mentally unwind, but avoid drinking alcohol as you need to be in peak performance for what lies ahead.

Typically, a route early in the second week of route selection will involve a 15-20km (9-12 mile) hill-and-mountain march carrying an 18kg (40lb) bergen and navigating between five or six RVs in a period of about four to five hours. By the end of the

week the distances will have risen dramatically to about 30km (19 miles), to complete in around six hours with eight or nine RVs. Gradients and the severity of the terrain to be tackled will also seem to be getting more demanding. This is a time when

the recruits are further thinned out, as they start to fall by the wayside in large groups:

'This is when it becomes a mental battle... you're getting a steady stream of people jacking [voluntarily dropping off] - injuries, attitude, and then of course you get those who simply can't move fast enough over the hills, 'cos the speed is crucial to everything. People are being rifted out, people are jacking, others aren't sure why they are there, they might have done Selection simply for getting time

Selection is focused on weeding out those who do not possess the right physical qualities, yet it is equally – if not more – concerned with discovering those who do not have the right character

off for training away from Germany. One or two who were there were just not all that fussed by it. I mean, this is why the hills are such a good test. You've really got to want to do it because it's f**king hard work. It's not a fitness test as such, I mean, you've got to be fit but you can't just be fit – because unless you're unbelievably fit and can just take it in your stride it's all about motivation.'

As this SAS soldier points out, selection is focused on weeding out those who do not possess the right physical qualities, yet it is equally – if not more – concerned with discovering those who do not have the right character to be a SAS soldier. The slightest hint of weakness or indecision on the part of a recruit during Selection will be picked up instantly by the DS. If it shows itself as a pattern, however slight, then that soldier will be RTU'd.

Sometimes the DS will play cruel jokes on the recruits to test their strength of character. One of the most common tests is when the recruit arrives at his final RV, exhausted both mentally and

LEFT: **The endurance march – the final exercise of 64km (40 miles) in 20 hours – sounds manageable at just over 3km/h (2mph), but add in the terrain, map-reading, 36kg (80lb) of kit, little sleep and chronic fatigue, and maintaining the speed becomes an incredible slog.**

Left: During winter Selection courses, give a priority to warm head-gear. Over 50 per cent of body heat is lost through the head, so keep it warm and dry. Draw a woollen cap over the tips of the ears, while a waterproof hood protects the face and will keep off rainwater.

physically, and the DS then tells him that the pick-up lorry is actually 10 miles away at a new RV and the recruit will have to march there. The lorry is actually only a couple of miles down the road, but the DS will watch intently for the recruit's response. If he suddenly seems to lose interest or patience, then he will probably be RTU'd. If he does not complain but sets off without a murmur, then he is behaving in the way the DS wants to see.

The third week of Selection accelerates the build-up to what is commonly known as 'test week'. In essence this part of the course is little different to what has preceded it. It is still navigation-based endurance exercises over the Welsh mountains. Yet the intensity of the courses are some of the hardest of any elite force around the world. Furthermore, they are conducted in total isolation – you stand or fall by your own decisions. The Brecon Beacons are a truly lonely place. Many recruits start to feel very depressed and disorientated at this stage. Their thoughts become more self-castigating and they can start to hear those voices that say 'go on, give up'. Indeed, those voices can be physically embodied in the DS. As the recruit arrives at each RV the DS will often say in a sympathetic and reasonable tone, something like 'Come on mate, you've given it your best shot, jump in the back of the truck, call it a day, and we'll take you back for some hot food and a shower.' If it looks like the recruit has to even think over this proposition, then he will be RTU'd. Complete motivation must be present from beginning to end, and anyone who doubts his own intentions will probably be a poor Special Forces soldier.

The endurance routes of test week are all well over 20km (12 miles), and have fearsome gradients. The routes will be specially chosen for their treacherous terrain and the difficulty in navigating them, and will demand that the soldier also show his aptitude at night navigation. The general weight of the bergen is now around 23kg (50lb), but in some exercises you'll find it more in the region of 32kg (70lb). In addition to this massive pack, you will have to carry your belt, rifle, water supplies, and an ammunition box full of stones or concrete which in itself will weight around 10kg (20lb). All in all your kit carry on such days may total about 41kg (90lb).

The foremost of all the Selection routes is the 'long drag' or 'endurance march'. As one SAS soldier says: 'Endurance – that's the big horror-story walk.' His expression is justified. Endurance kit will total about 36kg (80lb). The route distance is 64km (40 miles). The time to complete it is 20 hours. You will have already done an exhausting march the day before, and after almost no sleep you will set off at midnight into whatever weather conditions the Welsh mountains are throwing at you. If your motivation and determination are not 100 per cent as you set off, then your chances of passing are almost non-existent.

'Endurance' is the last phase of Selection. Pass this most formidable of tests – i.e. do it in the time allocated without missing an RV – and although you are not yet in the SAS, at least you will have made it through to Continuation Training. Although this is far from the end of the road, all those who have passed Selection will tell you that it creates a powerful sense of achievement and relief. From this point on you will start to learn the actual skills of the SAS soldier.

How to pass Selection

Now we have a detailed overview of the nature and content of Selection, it is time to backtrack somewhat and see what it is that makes some people fail the course and others pass, and what you could do to maximize your chances of passing.

The first thing to say straight away is that there are no short cuts, no tricks, to passing Selection. You certainly cannot cheat. Over the years the DSs of the SAS have seen all the attempted tricks, and know all the attempted short-cuts, everything from getting mates to pick you up in vehicles to having forbidden equipment buried at certain locations. Be sure of this: you will be caught and sent back to your unit. Then again, if you are the type of person who would cheat, even if you did manage to pass Selection your character flaw would eventually emerge in some other situation. Far better to pass squarely on your own merits.

In chapter 2, we looked into the techniques of navigation that you will need for Selection. The routes throughout Wales are where these skills come alive, but to maintain a good speed between RVs you need more than just compass and map skills. Firstly, when you have a grid location choose your route to it intelligently, not just on a crude 'as the crow flies' judgement. Sometimes the quickest routes might not be the most direct. Take into account the gradients indicated by the contour lines as well as any information revealed about the nature of the terrain.

Getting lost is perhaps the main reason for lost time – second to a failure of physical strength. Map reading and compass navigation must be carried out quickly and efficiently, and regular checking is the best method to prevent unwanted deviations from your route. Do not worry if the others taking the course are less circumspect and seem to be ploughing ahead with scant regard for the compass. Especially in bad weather with poor visibility, your caution will be rewarded with more accurate direction, though you should be sufficiently confident to take a reading and then trust it enough to push forward with all speed.

To keep your energy levels high while not wasting any time during Selection, drink while you are on the move (a drinking pipe connected to your water bottle on the belt is one of the best methods of doing this). Furthermore, be careful that you do not eat too heavily – you will be feeling nauseous enough from the exertion without a full stomach adding to your problems. Instead, eat from your rations little but often, just enough to keep giving yourself a much-needed energy boost as required.

However, perhaps the most significant problems you will face during Selection are psychological difficulties and health problems. We will first deal with psychological problems: defeatism, depression and a failure of determination are amongst the foremost reasons for failure. It is understandable to see why Selection should engender such dark moods, and

LEFT: Camping down for a short period of rest is one of the best ways to renew energy and restore morale. Use the time wisely – eat a meal, dry out any wet clothing, repair any damaged equipment, consult your maps to get a bearing on the next stage, and treat any injuries.

incline the recruit to give up even if his performance is adequate to requirements. SAS training is constantly tough and demanding, and the combination of poor weather, lack of comfort, physical pain, a critical DS, and a terrifying rate of drop-out can gang up on even the toughest of souls.

As the SAS soldier quoted earlier stated, many people who attend SAS Selection do so for the wrong reasons, and so write their failure before they even start. To be sure about the SAS as a vocation and to keep passionate about passing the training will be amongst your greatest advantages when attempting the Selection process. All those who attend Selection will reach the point of maximum exhaustion, but only those who have the grit and determination to push through the pain of fatigue and even injury will get through.

One important point about motivating yourself through the darker times is that you should return

You must ensure that you have a solid reason for attempting the Selection course in the first place. You must force this reason into your mind when you are flagging

again and again to why you are putting yourself through this (assuming that you know the answer to this question). Fatigue can make the world seem quite unreal, so firmly tell yourself – out loud is best as the sound of your own voice can snap you out of this introversion – that successfully completing Selection is necessary to get you into the world's most elite unit. Use any mental imagery you can to snap you into action. Imagine the pity and condescension of those back in your original unit if you were returned. Some SAS soldiers have said that during Selection when times were tough they imagined their children's faces, and were filled with the desire to make them proud of their dad. Whatever the case, it is best to decide beforehand what it is that is motivating you most strongly to pass Selection and bring this image to mind constantly and forcefully throughout the training.

One of the hardest issues to deal with for many is the problem of injury. Some injuries, such as blisters

and bruised heels, are common to almost every recruit who passes through SAS Selection. Unless, for example, a blister becomes so infected that there is a real danger of blood poisoning, then these are the type of injuries that will be treated with disdain by a medical orderly – you just have to keep going through the pain no matter how excruciating it is. The best way to deal with constant pain is to either ignore it or to defy it. Ignoring it is not a piece of facile advice. Try pretending that the pain is unimportant to you; do not reflect on it psychologically, but trivialize it in the way that you describe it to yourself in your inner voice. Do not think 'this is agony' but rather 'my body can handle this, no problem'. You might find that defying the pain is a better option. Get aggressive with yourself and push yourself even harder, challenging pain to do its worst. Being competitive can help here. One SAS soldier who describes his shin tendon during Selection as 'knackered', pushed himself through by summoning this sense of fiery competitiveness:

'... if I saw the bergen of the guy in front of me I'd try and catch him, and if I thought that anyone was trying to catch me, I'd make sure they didn't. So when the fog came in I'd have a bit of a problem 'cos I couldn't see the opposition, so to speak. I just switched off a bit, so I'd walk, and do the navigation, and then the fog would lift and I'd see someone and I'd think, "Right, let's go and get the b***ard", and off I'd go.'

Though depression and despondency during Selection are common to many, solutions to such feelings are individual. However, you must ensure that you have a solid reason for attempting the Selection course in the first place. You must force this reason back into your mind at moments when you are flagging. Remember the bit of Confucian wisdom – unhappiness results from the discrepancy between the way you want things to be and the way they actually are. Accept wholeheartedly the

RIGHT: Abseiling is a good way to shave some vital minutes off a punishing Selection schedule, though it is usually conducted when in pairs in case of accidents. However, do not attempt it in the dark or in very bad weather conditions.

way reality is and do not wish for anything else and you will gain strength. In the context of the SAS, this means accepting that the toughness of Selection is the way things are if you want to be the best.

FIRST AID FOR INJURIES

Injuries are frighteningly commonplace during SAS Selection training, indeed they are a common reason for being RTU'd. The vast majority of injuries are associated with the activity of constant walking under heavy loads over uneven terrain – blisters, swollen or twisted joints (especially ankles and knees) and occasionally broken limbs. Yet as we have already seen there are more life-threatening conditions connected specifically with climatic conditions. As Selection proceeds whatever the weather, the recruit faces the twin perils of heatstroke in the summer and hypothermia in the winter, both of which can have lethal consequences if left untended. In the remainder of this chapter we will look at how to deal with these types of climate-induced injury. We will describe the techniques as if you are delivering them to a third party. However, as so much of Selection is performed solo, you should also think about how you would apply the techniques to yourself, and recognize the symptoms should they arise in you.

Hypothermia and hyperthermia are opposite sides of the same coin. The human body, however much the external temperature varies, maintains its core temperature (the temperature in the central organs of the torso and the brain) at 36-38°C (96.8-100.4°F). In order to maintain this temperature, the body has a wide range of complex mechanisms for either retaining heat in the core (such as drawing blood away from the skin surface to slow heat loss, hence your fingers go cold in icy weather) or repelling heat (through the increased flow of blood to the skin surface and the increase in sweating). Disorders in this normally efficient system occur when the environmental conditions are so extreme that the body cannot cope and the core temperature goes beyond its safe levels. Hypothermia is the result of a fall below the core temperature and hyperthermia – more commonly known as heatstroke – occurs when the core temperature climbs above its normal range.

We will deal with hypothermia first. Those carrying out Selection during the winter months need to be acutely aware of this condition. It is most commonly suffered by those who are exposed to a combination of cold, wet and wind – generally the permanent conditions of the Brecon Beacons during the wintertime. Taken together, these elements will suck out body heat extremely efficiently until they start to impact on the core temperature. Once established, hypothermia is exceptionally dangerous, no matter how fit you may be. Remember that even the illustrious Bravo Two Zero patrol in the Gulf War suffered fatalities from hypothermia after

BELOW: The field treatment for hypothermia involves: a) Dig a snow trench and lay insulating material (such as leaves and branches) in the bottom; b) place the casualty in a survival bag; c) place the casualty in the snow trench and get in with him; d) hug him to share as much bodyheat as possible.

unexpectedly being caught within a blizzard in the exposed deserts of Iraq.

Preventing hypothermia involves following a number of essential measures. Cover your head and hands at all times with thermal hat and gloves respectively (remember to take a blue or black woolly hat with you to Hereford before starting Selection). Do not allow yourself to become soaked – use your waterproof outer clothing as soon as rain, sleet or snow sets in. Watch out if you are becoming extremely fatigued with a sense of detachment from reality; this is the time to have a rest period and something to eat and drink.

The problem with identifying hypothermia is that it can sneak up over a prolonged period. However, there is a set of common symptoms which you should look for in others or monitor in yourself if possible (you will be unlikely to have a medical thermometer with you on Selection, but if you do have access to one, then any reading taken from the mouth or groin below 36°C (96.8°F) should prompt immediate action). The person's mental state will be deteriorating at a rapid rate, with alternations between apathy and aggression, confusion, swings in energy level, as well as a general detachment from his surroundings. He will shiver violently and

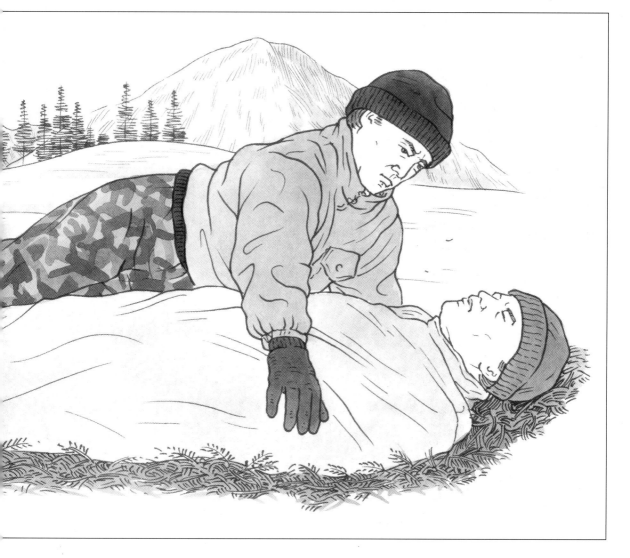

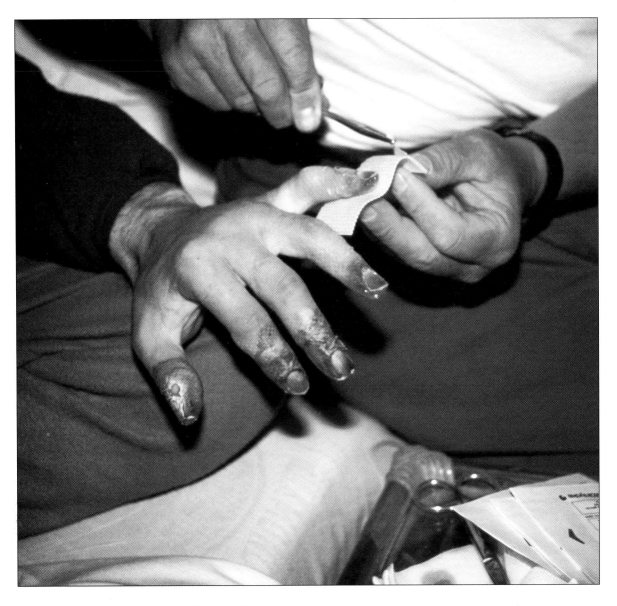

ABOVE: Frostbite is an excruciating condition which can result in lost limbs and nerve damage. Handle frostbitten fingers very gently and slowly bring them up to a normal temperature, preferably through sequenced immersion in warm water, or initiate rescue procedures.

have a very pale complexion, although this shivering can stop when hypothermia is well advanced – at this point casualties can often say they feel quite warm. His movements will also become clumsy and relatively uncoordinated.

You need to act very quickly and follow this simple system of treatment, otherwise the casualty could quite easily die:

● Immediately place the casualty in a position sheltered from the rain and the wind. If you cannot find a natural shelter dig a hole in the ground or snow and place him in it.

● Try to change him into dry clothing or at least protect him from getting wetter.

● Wrap him in his survival bag, and if possible place some insulating material between him and the ground (more heat is lost through contact with cold ground than cold air).

● If you can make a fire do so. Place the casualty near the heat or hug him in order to share your body heat (the procedure followed by the SAS corporal who attended to Kealy). Heating up stones, wrapping them in cloth and then putting them in the casualty's armpits and between his thighs is another effective treatment, though take care not to burn the casualty.

● Give him some warm food or drink if he is conscious and capable of swallowing.

● Do not handle him roughly – his major organs are weak and clumsy attempts to manhandle or carry him could result in cardiac arrest.

● Finally, using your flares or any other means, attract attention and initiate rescue proceedings.

While stuck on a mountainside, the above measures are all that you can really do to prevent the casualty deteriorating further, but they can be quite successful if applied properly. Yet before turning to hyperthermia, it is worth a quick look at another cold-related condition that could assail you on Selection – frostbite.

Frostbite results quite simply when the fluid content of parts of the body – usually exposed parts such as the fingers, toes, ears and nose – literally freezes solid. If your head, feet and hands are kept covered in appropriate winter thermal clothing, frostbite should not really occur on Selection. Yet all too often recruits discard items of clothing when they get hot and do not notice that the part of their body is slowly beginning to freeze. Frostbite is much more likely to occur in snowy and icy conditions combined with high winds. The skin of the affected part goes white and hard, then swells and slowly turns more towards a blue coloration. After that it can go black and infected. Intense care needs to be taken at any stage during frostbite. Imagine taking a thin frozen sausage out of the freezer and trying to bend it with force – a frozen finger can snap every bit as easily. Frostbite should only be treated with slow defrosting under controlled conditions, so it is best to simply wrap the infected area

and get the frostbite casualty to help. If you were to apply a full treatment, you would immerse the frozen part in comfortably warm water until it was thoroughly defrosted (keeping the warmth maintained by the regular input of hot water), then wrapping it gently in a bandage and administering anti-inflammatories. However, during Selection it would be very difficult to administer this type of treatment. You could also endanger the patient further by defrosting the body part and then allowing it to refreeze again because rescue has not yet arrived. The damage that results through refreezing is irreparable; the body tissue is literally shattered by the secondary expansion of ice crystals. It is far better to wait for the professionals to arrive and only contemplate those more advanced field treatments after you have completed the SAS medical course under Continuation Training.

No matter how exhausted you are, it is particularly important to look in control both physically and mentally when you meet the DSs at various RVs on your Selection routes

Hyperthermia may result from the opposite environmental conditions to hypothermia and frostbite, but it requires equally prompt action. Heatstroke usually results through excessive exposure to direct sunlight and heat. This increases sweating, and it is easy in very hot or physically exhausting conditions to sweat more fluid out than you are drinking. Once this occurs then the blood becomes thick, blood pressure falls, and dehydration sets in. These conditions are very serious indeed, and when advanced the chances of reversing them become slim.

Heatstroke is usually preceded by another condition commonly known as heat exhaustion. The symptoms of heat exhaustion include a fast pulse, mental confusion and disorientation, pale and clammy skin, cramps in the arms, legs and abdomen, nausea, as well as reduced urination. Immediately stop the casualty from exercising, and get him to sit down somewhere cool and sheltered from the sun. Give him small, frequent sips of water to increase his fluid levels (this can take several litres before he starts to feel better). Once his pallor and mental state start to recover, then (if you have not already done so) begin a medical evacuation. The overheated casualty might insist that he is strong enough to continue Selection – he is certainly not. Allowing him to continue could lead to the far more serious medical condition of heatstroke.

The heatstroke casualty will exhibit all the symptoms of heat exhaustion, as well as beginning to slip in and out of consciousness. His skin may go very dry and hot, and his temperature will have climbed over 40°C (104°F). His body temperature needs to be cooled rapidly. Do not throw him into a river or stream – the shock to his now fragile body system could kill him. Instead, follow the procedure developed by the Israeli Army, a force that knows better than most about operating in severe heat conditions:

● Strip the casualty to his underclothes.
● Spray or soak his body with water, then fan him rapidly. (Evaporation is the most efficient method of body cooling.)
● Once the external body temperature starts to lower, cover the casualty with a sheet soaked in water and keep the sheet regularly drenched.

Whatever your treatments, professional medical help is needed to control heatstroke, so the casualty must be placed in rescuers' hands as soon as possible. Heatstroke is not too common on Selection, because the air temperatures in the mountains tend to be fairly cool even in summer. However, heat exhaustion can prey on anyone doing the summer Selection (or winter for that matter if water intake is not kept high), especially when around 64km (40 miles) of exertion are exacerbated by high temperatures and an unobscured sun. To prevent the onset of heat exhaustion, drink plenty of water and keep your head covered at all times. Do not shed sweat-soaked clothing – it will keep you cooler than having your skin exposed and the sweat evaporating constantly. Finally, rest if you feel the onset of any disoriented state of mind, this is the signal that all is not well. Remember that even experienced SAS soldiers can be beaten by the ferocity of the climate.

We have now seen what Selection training consists of and its harsh demands on the SAS hopefuls. If you pass it, then you are through to the next stage – Continuation Training – in which you will begin to learn the combat and survival skills of the SAS soldier. Before we move on to this in our next chapter, it is worthwhile to take a moment's reflection on how you should behave during the Selection period. This will maximize your chances of gaining the approval of the DS and avoid being RTU'd.

Generally, avoid consciously trying to impress the DS – he will certainly not be. Instead, simply do everything he asks with a responsive and intelligent attitude. Show him that you are individually engaging with your task, not just following the crowd. Therefore, if you disagree with the navigation of another recruit during Selection, do not follow him, but stick to your own informed decision.

Stay 'switched on' at all times, and respond to unpleasant commands without a murmur of dissatisfaction. Particularly important points of the course are when you meet the DS at various RVs on Selection routes. No matter how exhausted you are, look in control both physically and mentally. Identify yourself briskly and efficiently. The DS will often give you a mental challenge at each rendezvous – such as a piece of mental arithmetic – and this you should perform confidently, without moaning about how difficult the task is. Words should be few and to the point. Throughout Selection do not pester the DS with lots of minor questions, as this will just annoy him and make you seem like someone who needs their hand holding. Conversely, do not be afraid to ask a question if it is important – it will show that you are thinking about what is ahead.

In a sense it is far better to allow yourself to merge into the background and let the DS notice you himself. If you highlight yourself through your actions then you are much more vulnerable to the scrutiny of the DS. Remember, if you do everything that is asked of you, that will usually be sufficient to get you through Selection. One final point regarding your rifle: SAS soldiers are fast-response troops who on operations have their weapons at the ready at all times. Never go more than arm's length away from your rifle at any time, and always make sure that it

SURVIVAL SHELTERS

For the purposes of Selection, it is not necessary to know how to build an advanced shelter. Yet it is advisable to know in advance how to construct a basic survival shelter as a precaution should route selection take a turn towards life-threatening conditions. The Brecon Beacons, though barren in many parts, still have enough natural materials around to construct a shelter substantial enough to protect those inside from severe wind chill and even rain.

The most instantly accessible form of shelter is that which is 'ready built'. A large depression in the ground, a cave, or a tree with a broken trunk forming a natural roof at the angle can all serve well as a temporary shelter. However, what is much better is to take a natural feature and then improve on its sheltering properties artificially. The basic principle is to take a recess or sheltered space and then construct further sheltering features around it to complete the protection. One of the easiest shelters to make uses the exposed roots of a fallen tree. Using tree branches (conifer branches are particularly good) make an angled roof from the ground to the top of the exposed root. Keeping the branches densely packed and tied if possible, use leaves and tightly bunched grasses to make thatching for the roof.

Repeat the process until you are protected on all sides (you can use your bergen as a door). This sort of shelter can be made against many types of surface, including bushes, rock surfaces and dry stone walls.

If the ground is covered with heavy snow and there is little natural shelter, a snow trench is probably your best immediate response. Cut out a trench in the snow just longer than your own body length and about 60cm (2ft) deep. If the snow has been cut out in firm blocks, use these to make an angled roof over the trench, with two sets of blocks meeting in the middle and holding each other up by weight. Otherwise, use branches to construct the roof. Then enter the shelter from one end, first lining the floor of the shelter with equipment to prevent you from coming into contact with the cold snow. Make sure your locator beacon is activated and your luminous bergen patch is outside, then huddle up inside your survival bag and wait until it is safe to move or rescue comes.

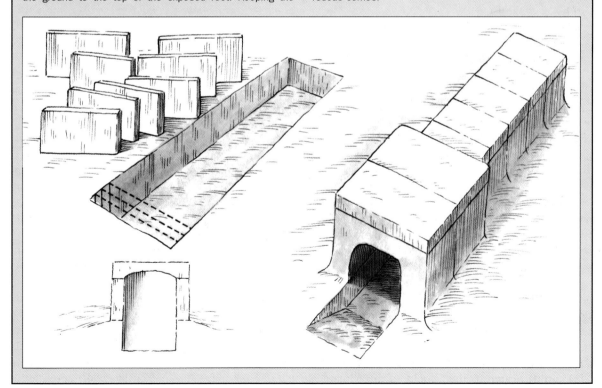

BLISTERS AND CARE OF THE FEET

Blisters are actually the result of friction burns. Heat builds up over a patch of skin which is being rubbed until a bubble of liquid-filled flesh forms to act as a sealed protection over the injury. As with any injury, prevention is better than cure. As soon as you feel a hot-spot building – it will usually be on your feet or ankles – change your socks (you should have one or two spare pairs within easy reach in your pack). Also, modern climbing suppliers offer a range of ointments and rehydration plasters which can be useful in controlling the onset of blistering. Use them as soon as there is any indication that blisters are forming.

Unfortunately, even with good preventative measures and modern treatments the severity of the Selection course makes it likely that blisters will form. The basic principle with treating blisters is to cover them with a plaster and leave them alone. Blisters form for a purpose - they provide an air-tight bubble over an injury which keeps out infection. As soon as they are opened, you run the risk of dirt and bacteria penetrating the wound and the blister turning septic. However, Selection tends to produce blisters of such grotesque magnitude that they start to impair the mobility of the feet. In this case they need to be burst – or rather lanced – to restore movement.

To lance a blister, take a sterilized needle (sterilize it by immersing it in alcohol, boiling it in water for about 10 minutes, or holding it over a flame – though you could also take pre-packed medical needles) and pierce the blister once at its lowermost point. Then use a piece of bandage to gently squeeze out the liquid contents. Do not pull off the skin covering. Instead, place a sticking plaster over the wound and then leave well alone.

Blisters are not the only problem which might assail your feet on Selection. Another trial is swollen feet. At the end of each punishing day, you might find that the sheer pressure your feet have been under makes them swell to sometimes alarming proportions. There is little you can do about this

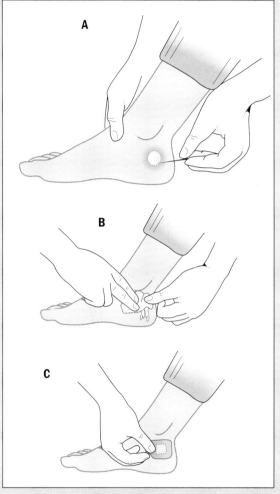

A

B

C

while on the move. Yet when you stop at night apply cloths soaked in cold water for some time to take down the initial swelling and then sleep with your feet propped up higher than your head. This will reduce the blood flow to the feet and ankles, and inhibit further swelling.

is clean and well maintained and ready for inspection. SAS legend tells the story of a recruit who left his rifle several yards away while he went to collect water from a lake. His punishment was 20 push-ups – at the bottom of the lake!

We always return to the point we made in the first chapter. The SAS do not want automatons, but require thinking, hard-working dynamic individuals who respond well to orders but have the practical intelligence to interpret those orders most efficiently. If you get through Selection, then you will have even more opportunity to demonstrate these important survival qualities during Continuation Training.

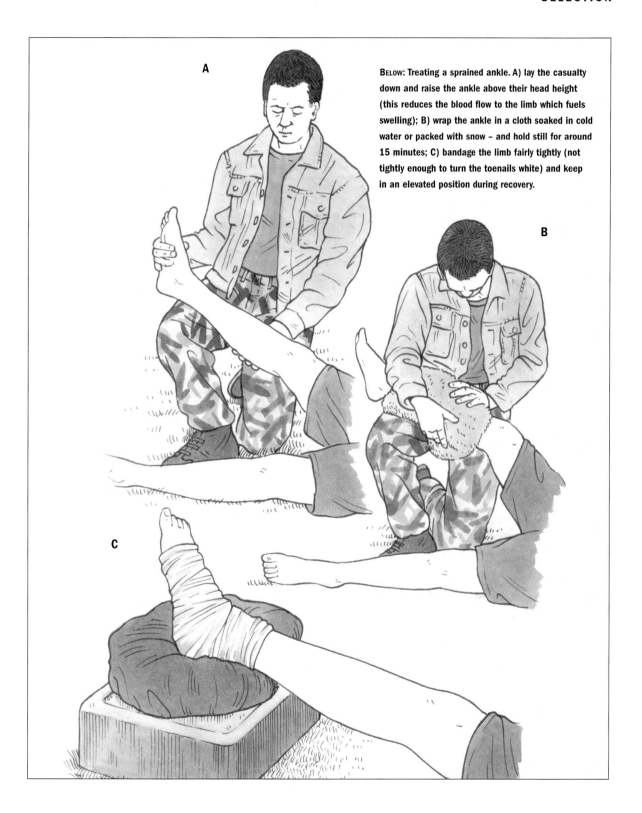

A

B

C

BELOW: Treating a sprained ankle. A) lay the casualty down and raise the ankle above their head height (this reduces the blood flow to the limb which fuels swelling); B) wrap the ankle in a cloth soaked in cold water or packed with snow – and hold still for around 15 minutes; C) bandage the limb fairly tightly (not tightly enough to turn the toenails white) and keep in an elevated position during recovery.

CHAPTER 5

Continuation Training

Continuation Training is a period of intense learning for the recruit. New skills come thick and fast, he fires weapons he has never even touched before, and he begins to learn those combat drills which separate the SAS from the regular forces. Yet as with earlier process of Selection, he can also fail at any time.

Continuation Training (CT) heralds the moment when the recruit actually starts to feel part of the Regiment itself. Selection has done its work. After a tortuous period of physical and navigational exercises, the Regiment is left with a group of around 10 men who have proved that they possess character, durability and tenacity, plus essential skills in finding their way from A to B. Consequently, CT switches its emphasis from testing the capacity for endurance to testing the capacity to acquire the skills of the SAS soldier.

CT is an exciting period for any recruit. He starts to visualize the beige beret as being his, the training is mentally stimulating and action-packed, and he is initiated into more of the secrets and lifestyle of this previously opaque unit. But caution is the watchword. Despite having come through the hardships of Selection, the recruit can still be dropped at any time over the 14-week period to follow – as the DS

LEFT: A trooper takes aim with an M16A2 assault rifle fitted with an M203 grenade launcher beneath the barrel. This is one of the many weapons which the recruit encounters during Continuation Training as he becomes acquainted with the broad sweep of the world's firearms.

105

ABOVE: The recruit will be used to the standard issue SA80 rifle, but during CT this is swapped for the M16, the SAS's preferred standard firearm. It is a 5.56mm rifle with a muzzle velocity of 1000mps (3280fps), the high-velocity round imparting terrible tissue damage.

will be quick to point out. (You must even bear in mind that if you are finally accepted into the Regiment you will still have to fulfil a probationary period.) As with any period of SAS training, the recruit must give 100 per cent from the first moment he wakes up to lights out late at night. If anything, there is perhaps more scope to slip up, such is the huge volume of new information and new skills which the recruit will be expected to absorb and apply.

OVERVIEW OF CONTINUATION TRAINING

Continuation Training, like Selection, shifts and changes in emphasis and content every year. The basic purpose of CT is to start the long process of changing British Army or Royal Marine soldiers into Special Forces soldiers. Many of those who pass into CT will already be good infantrymen or Marines, but CT must impart the fundamentals of being an elite operative, and teach Regimental specialisms which the recruit is unlikely to have encountered before. In addition, CT also continues the process of assessment that began in Selection, so CT remains a testing process as much as a learning one. Yet here lies the real incentive – if you can

pass through the next 14 weeks of CT, then you are in. You will receive the SAS beret and you will then enter into one of the SAS operational units, Sabre Squadrons. The recruit must use this strong incentive to propel himself through CT as there are some physically tough times ahead, and some uniquely harrowing challenges.

CT is a 14-week period which contains instruction in the following elements. Firstly, the SAS's Standard Operating Procedures (SOP) are taught. This means the basic combat tactics fundmental to the SAS such as the four-man patrol, covert manoeuvres, reconnaissance techniques, insertion methods, and contact drills. This section of the course is always an exciting one for the recruit. Not only does he practise fast-moving live-firing drills, he also uses a range of weaponry. This will include foreign weaponry, as the SAS soldier must essentially be familiar with the firearms of any nation he might find himself serving with or against in future operations. In addition, the recruit will receive training in demolitions and the handling of various combat vehicles, and he will get to 'have a go' in the Regiment's infamous hostage-rescue scenario, the so-called 'Killing House'.

Part of the SOP training also includes instruction in signalling. SAS recruits are obliged to meet the British Army's Regimental Signaller standard, and they will be versed in non-standard techniques of encryption and covert communications. The recruit will also find out about more sophisticated methods of navigation than those employed during the Selection process. In particular this means using the Global Satellite Positioning System (GPS), which has a very high degree of accuracy and speed. Other elements of the first phase of CT include basic combat first aid.

The next section of CT is the Combat and Survival Training, and here the course starts to get physically tough once again. Combat and Survival Training focuses on the skills of staying alive in environments which are hostile both physically and militarily. This part of the course is one many SAS soldiers have had cause to lean on in the reality of their operational lives. SAS soldiers tend to operate far from help, often deep behind enemy lines. Once their rations, ammunition and even mission are exhausted, then they must dig deep into their own resources to stay alive. Combat and Survival Training runs through the building blocks of survival methods, from finding food and water to making fire and surviving natural disasters.

The survival training which takes place in the UK naturally focuses on survival in temperate zones. However, CT also features a much more specialized environment when the recruits undergo up to six weeks of jungle-survival training in Brunei at the British Army's Jungle Warfare School. This is a vivid and exotic period of the CT, and not an easy one. For some the plunge into the jungle can be disorientating, for others it can be fascinating. Yet whether in the jungle or back in the UK, the training will place recruits in positions of genuine survival. Many times they will find themselves alone in unforgiving environments with only the clothes they stand in, and have to stay alive using the techniques they have been taught.

Yet the survival phase of CT is perhaps not the worst element of the 14-week course. Taking place in the UK, Escape and Evasion signals that the recruit is nearing the end of CT. The details of this phase are explored in the next chapter, but the final element involves a pursuit of the escaping recruit followed (when he is captured) by a simulated interrogation under the control of the Joint Services Interrogation Unit (JSIU). Although simulated, the interrogation is conducted with a degree of brutality and realism, and the DS will be looking for any weakening of resolve on the part of the recruit undergoing this painful experience.

After Escape and Evasion, the recruit reaches the final stage of Continuation Training, when he will have to undergo up to four weeks of parachute training at No 1 Parachute Training School at RAF Brize Norton (though the sequence of these two parts of the course might be reversed on occasions). This is where it can be an advantage if you are trying to get into the SAS from The Parachute Regiment, as you will already have completed a parachute course at Brize Norton. The place and the sequence of events will therefore be familiar to you, but approach the training with all the seriousness you would if you were doing it for the first time. For the SAS parachute course is actually far more intensive than that of The Parachute Regiment. The training is focused exclusively on static-line para-chuting (training in more advanced parachute techniques comes once you have entered the Regiment), and you will have to make about eight successful jumps – including night jumps and jumps into water – before passing the course.

There are many more elements to CT, including tests in language skills and engineering ability, and there is doubtless much about CT which is not in the public domain. The course is consistently intense, as CT searches out your intelligence more than your stamina. Nevertheless, if the recruit has managed to complete the 14 weeks of CT, once he has made his final parachute jump he steps forward to become a full member of the SAS, an awesome moment by the accounts of those who have completed the full process of SAS Selection. In this chapter we will look in particular at the combat

elements of CT, leaving the survival and interrogation aspects to a chapter of their own.

WEAPONS TRAINING

Small arms training is one of the most exciting aspects of CT for the recruit fresh from Selection, as an NCO of 22 SAS here admits:

'Yeah, it was a real eye-opener. Being a sapper I'd fired the SLR, the SMG, the GPMG and the LMG, and that was it: bust. I'd never even touched a pistol, let alone fired one. And even when we did go on the ranges, which wasn't that often, you wouldn't get to fire that many rounds. It was always a case of twenty or so to check the zero and then fire your APWT, or some buckshee shoot. So then we started on the M16, and the M79, and the M203, and the Minimi, and the Kalashnikovs, and the foreign weapons, and we're using grenades and mortars, and pistols: it was what I'd joined for; soldier heaven. I thought: "This is me!"'

The excitement behind this soldier's experience of firing such a diversity of weapons belies the methodical and serious process at work. As we have already noted, SAS operations often take the soldier far beyond the logistic support of British forces. In terms of ammunition, they can take only what they can carry. As experience shows, even fire-disciplined soldiers such as the SAS will expend much of their ammunition supply in a significant encounter. Once the ammo has gone, then the soldier must use whatever arms are to hand, and thus he must be trained properly in how to use the world's firearms. Consequently, the SAS recruit passing through CT can expect to handle something in the region of 30-50 individual weapons.

The first weapon the SAS recruit is likely to encounter is the US M16A2 assault rifle. This is actually the standard rifle of the SAS, having rejected the British Army's SA80 on the grounds of poor reliability. The M16 is a 5.56mm high-velocity weapon, gas-operated, and feeding from a 30-round magazine. Properly maintained, the M16 will give consistent, accurate performance, and its 1000mps muzzle velocity has a potent knockdown power through the effects of hydrostatic shock (the process

PRINCIPLES OF EFFECTIVE SMALL ARMS HANDLING

● Have a target to shoot at before pulling the trigger – do not just spray in a particular direction unless you are providing covering fire.

● Use semi-automatic mode at long ranges (which has more accurate fire), utilizing three-round burst at closer ranges. Only use full-automatic if faced by a mass assault where hits are ensured.

● Squeeze the trigger, do not snatch it – this disrupts the stability of the aim.

● Include two tracer rounds at the bottom of each magazine – these will tell you when you are about to run out of rounds.

● When going into a situation when you will need constant firepower, change to a full magazine even if the old one is not fully expended.

● Do not occupy a single firing position for too long if possible – it will attract increasing fire. Instead try to shoot and move, either advancing upon or withdrawing from the enemy.

● When picking targets, try to hit officers and so break the command structure and the discipline of the enemy units attacked.

● Do not shoot the enemy soldier just once or twice. Continue to fire until he goes down.

ABOVE: The Sig-Sauer P226 is much used as an SAS side arm. Sig-Sauer pistols are comparatively expensive for military weapons, but their reliability is extraordinary. It was destined to replace the Colt M1911 as the US Army standard pistol in the 1980s, but was beaten by the Beretta 92 on price alone.

whereby the water-content of human tissue 'ripples' under the supersonic shock wave of the bullet passing through it). The M16 also comes with a separate grenade-launching unit, the M203. This attachment is fitted just beneath the barrel with its own trigger unit, and is breech loaded with a 40mm grenade (ejection is provided through a pump-action mechanism which wraps around the launch tube). The M203 is controllable (it can be launched from the shoulder) and also accurate up to 400m (1312ft). With anti-personnel and anti-armour warheads, the M203 enables the SAS soldier to launch a fairly devastating attack from his assault rifle alone, without having to change between weapons.

The M203 replaced the M79 grenade launcher, a separate weapon in its own right. The M79 has similar performance to the M203, though it is far heavier and – as it is a separate gun – far less convenient. You will still receive training in this weapon as an SAS soldier. One of the criteria governing which weapons you are trained in is whether they are still in widespread use throughout the world. The M79 'Blooper', for example, was used massively during

If the recruit completes the 14 weeks of Continuation Training, he steps forward to become a full member of the SAS, an awesome moment

the Vietnam War, with the consequence that when the US withdrew from that conflict in the late 1960s and early 1970s many thousands were left behind. They ended up in the posession of various civilian and military factions right across southeast Asia. Thus with the SAS's track record of operations in that region, the M79 fits appropriately into the SAS training course.

However, for the moment we return to the weapons which form part of the standard SAS 'kit'. For counter-terrorist and urban combat use, the most easily identifiable weapon is the Heckler & Koch MP5 sub-machine gun. The MP5 deserves its reputation for excellence, and was selected by the SAS for its exemplary quality and accuracy. Like most sub-machine guns it fires the ubiquitous 9mm Parabellum round. Yet unlike many sub-machine guns, it fires from a closed bolt. Open-bolt guns propel the entire bolt system forward upon firing, the resulting shift of mass destabilizes the user's hold on the gun, and consequently affects the accuracy.

The MP5 is different. Its bolt has already locked in place prior to the trigger being pulled, with the result that only the firing pin shifts forward and so accuracy is maintained at all times. This is essential when considering the MP5's main use within the SAS: counter-terrorism and hostage rescue. As most of these types of operations take place in confined buildings with non-enemy personnel present, accuracy must not be compromised. The MP5's 800rpm also ensures a rate of fire sufficient to counter even the most fanatical of terrorists.

Unlike most other British Army soldiers, the SAS soldier also tends to carry a pistol with him. This could be either the Browning Hi-Power or, more likely, its replacement the SIG-Sauer P226. Both are excellent 9mm weapons. The Browning served the SAS well for many years and doubtless it will appear in SAS hands for many years to come. However, with its 15-round magazine (the Browning has 13-rounds), ambidextrous magazine release, and exceptional quality of production, the P226 has key

SAS HAND GRENADES

The SAS has a much greater budget per capita for ammunition expenditure than regular units of the army, thus the recruit entering Continuation Training will find himself able to practise the use of hand grenades much more frequently than other recruits in the past.

The most dominant type of hand grenade used by the SAS is the L2A2 anti-personnel fragmentation grenade, the standard-issue grenade of the British Army. This contains 170g (6oz) of explosive and weighs 450g (1lb). It has a four-second delay and will injure and kill up to 20m (65ft), most of the lethality coming from splinters from a pre-split coil of wire which is wrapped between the outer casing and the explosive. A range of other hand grenades are produced by the British firm Haley & Weller which have similar properties to the L2A2, though with the crucial difference that they have electronic fuses which are totally silent, excellent for use in types of ambush scenarios.

advantages, and the recruit will train on this weapon until he has good control over a 50m (164ft) range.

In terms of machine-gun technology, the SAS use the standard British Army tools of the 7.62mm General Purpose Machine Gun (GPMG) and Browning .50 calibre (when they are using vehicle mounts). However, because much of the Special Air Service's work is carried out on foot, and wielding a lengthy machine gun is almost impossible on covert missions, the SAS also tend to utilize 5.56mm light machine guns. Most popular amongst these is the US Army's Minimi. The Minimi is quite a flexible weapon in that it can be belt-fed from a 200-round belt like most machine guns, but it also accepts the M16 standard magazines. Importantly, switching between the two requires no modification in the field, just a simple switch between one feed aperture and the other. Its weight is a very reasonable 6.85kg (15.1lb) - compare that with the GPMG weighing in at 10.15kg (22.4lb) - and its length of barrel means that accurate suppressing fire can be put out to up to 800m (2625ft) at a rate of fire of 700-1000rpm.

Thus run the standard firearms of the SAS, but the training in firearms that you will encounter during CT extends well beyond the remit of standard-issue Western military weapons. During the first weeks of CT the recruit will get to handle a bewildering array of firearms. One weapon with which he will have to become intimately acquainted is the Kalashnikov AK47 and its derivatives. Since it was introduced as a standard Soviet infantry rifle in 1948, the AK47 has gone on to be the most prolifically produced rifle in history. Including all the international variants of this gun, it is estimated that that some 80 million have been produced. The fact that the AK is almost indestructible means that most of these are still in service. The Cold War saw to it that AKs were distributed around the globe, though today a large percentage of the 80 million weapons are unaccounted

Whereas regular troops can often disintegrate into a 'spray and pray' approach, an SAS soldier does not have that luxury, so must make every bullet count

for and exist in the hands of terrorists, factions, minor armies and ordinary civilians in most countries outside of Europe. In Africa, for example, AKs are so common that they can be bought for about $6 or even swapped for a bag of maize or rice.

Though the AK is far less sophisticated than the M16, it is built like a tank and simple to operate. Bury an M16 in snow or mud and then dig it up, and it will need some careful cleaning before it is safe to fire again. Do the same with an AK, and all you'll need to do is put the selector to fire and pull the trigger. The AK is simplicity itself to operate and field strip, and the SAS soldier will have to become almost as familiar with this gun as his issued weapons. Other weapons which have a similar ubiquitous distribution are the FN FAL (in the British Army known as the SLR, and once the standard issue British rifle) and the Heckler & Koch G3, which still kit out many armies worldwide.

Naturally, given the time available to the recruit during CT, he cannot become expert in every type of firearm which is used around the world. What the

DS is looking for is someone who can load, fire and maintain the world's most popular weapons, and consequently be able to understand the principles of weaponry enough to operate an unfamiliar gun in a combat situation.

We must also look at what is distinctive about SAS weapons training in terms of actual shooting skills. Special Forces invest huge amounts of time and ammunition in ensuring that their soldiers are excellent combat marksmen. An example from recent Special Air Service history clarifies why. During the Gulf War, the noted eight-man Bravo Two Zero unit came into contact with an Iraqi force many times greater in number. During the resulting battle – in which the SAS unit suffered no casualties whatsoever – the Iraqis suffered such a high rate of dead and wounded that the commander later reported that he had been attacked by a force numbering in the hundreds.

The difference between the SAS and the Iraqis in this particular incident was that the SAS are trained to control their fire, select a target, accurately aim at it, shoot until it goes down, then switch to the next target and simply repeat the process. Whereas regular troops can often disintegrate into a 'spray

ABOVE: Stun grenades are a key part of the SAS arsenal for hostage-rescue missions. They are a non-lethal munition (there are few metal parts to create lethal fragments), but produce 160db of noise and a blinding flash of 300,000cd, which combine to render an opponent momentarily incapacitated.

and pray' approach because they have the support of mass-troop firepower and lots of ammunition, the SAS soldier does not have that luxury, so must make every bullet count. Furthermore, scientific research has demonstrated that less than 15 per cent of regular troops actually fire their weapons in combat, such is the paralysis that takes over during the stress of action. Conversely, SAS soldiers work in such small units that everyone must fire and fire well once the action starts. That way they can have an impact on the combat well beyond their unit size.

CT training will want to see the recruit demonstrating these skills. His firing posture must be balanced and solid, and flexible enough to move rapidly to a new firing position. Sighting should be accurate and fast. Trigger pull must be done quickly but without jerking or snatching. Controlled bursts need to be put into a meaningful part of the target – generally the torso as this presents the broadest body mass and contains many vital organs – with controlled recovery between each burst, and then the next target ruthlessly sought out.

Essentially, three styles of shooting will be taught to the recruit during CT. The first is rapid-response firing, particularly useful when faced by an ambush. Here the front sight only is aimed towards the target with a quick glancing action – there is no time to align the front and back sights – the very top of the sight placed slightly lower than the desired point of impact (there is a tendency to shoot high during quick-fire encounters). This type of shooting has relevance up to about 15m (49ft). Beyond that distance then the soldier will have to employ rapid aimed fire. The crucial feature about this is recovery between shots. The soldier should let the barrel of the weapon fall back into the starting position as naturally and swiftly as possible. It is easier to do this with some guns than with others. A modern assault rifle such as the 5.45mm Russian AK74 can be fired on full-automatic and still produce tight groups at up to 100m (287ft), such is the control of gun climb through a very effective muzzle brake.

The third type of shooting is the skill of night firing. This is particularly difficult because the human eye functions in a different way in the dark. In daytime vision, light is projected squarely onto the light-sensitive cones behind the retina. However, in darkness it is the optical rods around the periphery of the eye which are more sensitive, though they take time to become accustomed to the darkness. Consequently, most soldiers tend to shoot high at night because they look at the centre of the target, yet the image enters their eyes at an angle to the normal line of sight. To compensate, elite soldiers

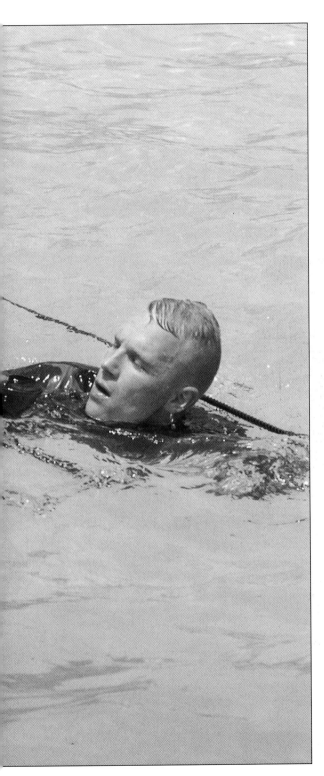

train themselves to look at the base of the target and shoot low, trusting that they are actually on target. This takes practice and repetition until the soldier trusts the process. The necessary habits of night-shooting will be built up throughout CT.

Of course, small arms are just one element of the weapons used by an SAS unit. CT training also includes a comprehensive introduction to the world's infantry-operated anti-armour, anti-personnel, and anti-aircraft systems. Of the anti-armour weapons, CT will usually focus on shoulder- or vehicle-launched weapons with effective ranges between 300m and 2000m (984ft and 6561ft). Occupying the lower ends of the range is the Russian RPG7, probably the most common anti-tank rocket system in the world, but with a compressed maximum range of about 400m (1312ft). Next come the one-shot disposable LAW 80 and M72 rocket launchers, able to reach out to about 500m and 1000m (1640ft and 3280ft) respectively. The M72 in particular is very popular as it is compact and light-weight, and easy to carry by patrol members. Finally, there is the MILAN missile system. The range of this system reaches out to 2000m (6561ft). Once launched the missile is guided to the target simply by the operator maintaining the target image in his sight – the information is fed to the missile through a trailing wire. MILAN is not common in SAS foot patrols, but during the Gulf War SAS combat and patrol jeeps had MILANs fitted to their roll bars and subsequently were able to take on a variety of Iraqi armoured vehicles.

There are many more weapons in the SAS arsenal. However, time limits during CT mean that further specialized weaponry training has to be reserved for when a recruit actually enters an operational SAS squadron. Yet it is no good to the SAS if the recruit learns to fire weapons but cannot employ them constructively in a coordinated combat formation or a pressurized combat situation. This is why the SOPs taught during CT will introduce recruits to a whole new way of tactical thinking.

PATROL TECHNIQUE

Perhaps the cornerstone of the SAS SOPs is the four-man patrol. All elite units have experimented with different sizes of patrol unit, searching for the most effective mixture of mobility, load-carrying, all-round surveillance, mix of specialisms, and fire-power. The SAS defined the four-man patrol as their

BELOW: Surrounded by three five-man squads – each containing a machine gun and anti-armour weapon – the central command group would trigger an ambush. The other groups would close the trap on all escape routes. The three-man team at the bottom is a guard element.

RIGHT: A typical patrol formation of a four-man SAS unit. Each man has responsibility for a dedicated field-of-fire, adding up to 360° fire coverage overall. The front elements take charge of navigation and route assessment, with radio operator and machine gunner at the rear.

ideal formation at the point of their reformation in the 1950s, and it has stayed ever since. Patrol members of regular SAS units will tend to be defined by a certain specialism, such as combat medicine, demolitions, and reconnaissance. The specialisms add up to produce a formidable fighting unit, highly capable and able to exhibit great independence

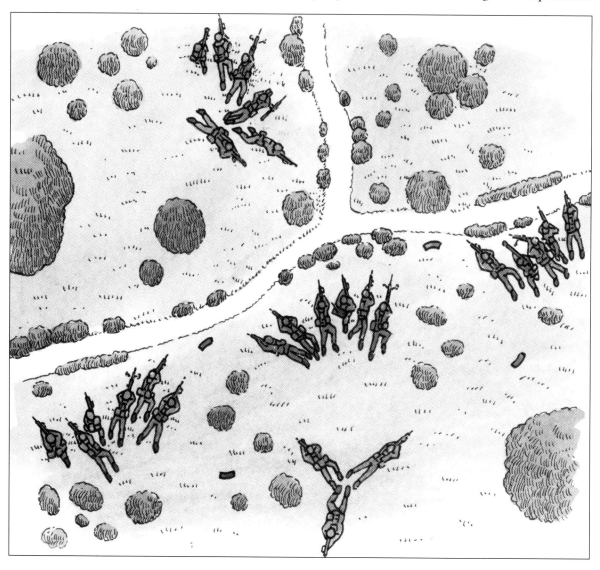

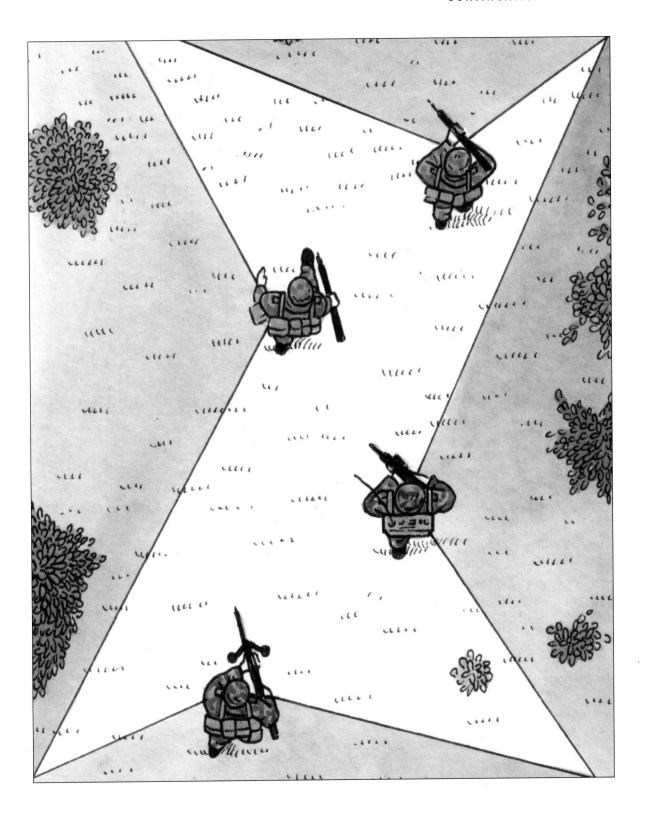

within whatever situation they face. However, as the specialisms are not imparted until the recruit actually becomes an SAS soldier, during CT the instructors are looking for recruits who can take their place in the patrol structure and contribute to it fully without forming a weak link in the structure.

Usually, SAS four-man patrols arrange themselves in single-file, box, or diamond formations according to the terrain and mission, and the patrol should be able to slip between these formations effortlessly and silently. As a general rule, however, each patrol will have a lead scout who acts as the primary eyes and ears of the patrol, including leading the navigation. He will also be responsible for spying any booby traps which may be placed on the route ahead, a job of maximum responsibility requiring strong nerves. Following him will be the patrol

A four-man patrol should add up to a 360-degree field of fire, with each man responsible for a designated arc of fire and surveillance

commander, checking the navigation and giving instructions forward and backwards through the patrol. Next comes the signaller, then at the end comes the second-in-command who has the job of protecting the patrol rear and also watching how the patrol members ahead are proceeding.

A four-man patrol should add up to a 360-degree field of fire, with each man responsible for a designated arc of fire and surveillance. CT training will test the patrol's ability to respond in any direction rapidly and in a coordinated manner. In ambush, they should be able to hit the enemy hard and ferociously for a few seconds, before dispersing in such a manner as to create severe confusion on the part of any of those left alive who might consider pursuit. Each member of the patrol should demonstrate awareness of the right fields of fire to create an effi-

RIGHT: Two SAS troopers quickly enter a window during hostage-rescue training. Such an entry would usually be preceded by a stun munition to distract and disable the opponent temporarily. Immediately inside the troopers will each clear a pre-agreed portion of the room.

cient ambush, plus exhibit excellent non-verbal and verbal communication skills to ensure the trap is set efficiently between the patrol members.

CT training will also introduce the recruit to various Contact Drills designed to maximize a strong response in case the patrol itself is attacked. Contact responses are practised until they become second nature – if they do not become so then the recruit will be given only a few chances before he is RTU'd. The DS will want to see the patrol act as one and fan out into defensive positions. During this fanning procedure one member of the troop – usually the lead man – should be opening up with his weapon and spraying the enemy positions. In only seconds the entire unit should be flanked out left and right and saturating the attackers' positions with heavy fire, even advancing upon the enemy to throw them into confusion. Upon withdrawal, a two-man covering element usually remains behind

The walls of the building are treated with a special rubberized surface which soaks up the thousands of live rounds fired in the Killing House every year

to hit any pursuers, but then these withdraw under the covering fire of the other two men who will have retreated a distance, but then adopted covering firing positions. Another withdrawal tactic is simply for each man to head off in a different direction with the agreement that they will meet at a designated rendezvous. As the patrol splits the pursuers will be left in confusion about which way to go and their resources will be split.

In theory, these patrol skills sound fairly clean and intelligible. Yet in practice, and under the confusion of gunfire, explosions and enemy counter-measures, they can disintegrate into chaos if each unit member is not highly disciplined. Thus the instructors during CT will throw up as many different scenarios as possible and see how the patrol members react. What they are looking for is the person who will retain presence of mind and rational thought even during moments of unbelievable pressure. The ideal scenario for testing this responsiveness is in the infamous 'Killing House'. From the outside the

Killing House – situated at Stirling Lines – does not appear to do justice to its name. It is a bland and windowless rectangular building of unassuming appearance. Yet it is here that the SAS train in the skills of hostage-rescue and counter-terrorism in an urban setting. During CT, the recruits will get their first taste of the high-speed tactics and responses necessary for dealing with a hostage crisis where seconds wasted might mean the difference between life and death for both hostage and rescuer alike. The full skills of hostage rescue are refined when the recruit has made it through to be a full member of the SAS.

The building is basically one long corridor with rooms on every side. The walls of the building – heavily constructed with massive metal-reinforced wooden sleepers – are treated with a special rubberized surface which soaks up the impact of the many thousands of live rounds fired in the Killing House every year. The rooms either side have no set configuration. The layout is changed constantly to give the soldiers new challanges as they head down the corridor, clearing every room as they go. Such is the flexibility of the design that the rooms can even be configured to look exactly like the interior of, say, a supermarket, train carriage, or aircraft.

The aim of the drills in the Killing House is to enter at one end of the building and, working as a team, clear each room of terrorists while rescuing the hostages. The whole process is designed to put the recuits under as much strain as possible. Imagine the enormous crash and flash of stun grenades, the roar of SMG fire in confined spaces, smoke filling the air, the pressure not to shoot an innocent hostage even when adrenaline is pumping hard – it takes rare self-control to move through the Killing House successfully. Real SAS members often take the place of the cardboard cut-out hostages. They must sit entirely still while stun grenades and 9mm bullets detonate around them. Each DS is primarily looking for someone who will maintain 'tempo', by keeping up the speed of the assault without any flagging moments of indecision or confusion. Rooms have to be entered and exited with clarity of purpose, and speed-reaction skills must be finely honed. Naturally, the first few excursions through the Killing House will be laborious and over-analysed by the recruits, but they will be expected to make quick progress and learn the proper assault procedures rapidly.

By constantly rehearsing these exercises, the procedures for an effective hostage-rescue mission start to become second nature, with the advantage of a fast, conditioned response rather than a slow, conscious thought-process. This vital learning process is helped by the use of closed circuit cameras to record the action within each room – this is then played back to the recruit so that he can see his mistakes.

BELOW: A sniper takes aim during training. His camouflage is known throughout the British Army as the 'ghillie suit'. Made from strips of green and brown hessian, it breaks up the outline of the soldier when set against foliage. Even the rifle is wrapped in strips of cloth.

We shall return to the Killing House in Chapter 6, to look at what is involved in creating and maintaining the elite hostage-rescue skills of the Counter Revolutionary Warfare Wing. However, now we turn to look at a less ostentatious part of CT – training in the arts of covert surveillance and deployment.

THE ART OF STAYING HIDDEN

It is an irony of the SAS that a unit that is so adept at combat skills will generally spend much of its time trying at all costs to avoid combat. When operating behind enemy lines, the numerical supremacy of the enemy will always be a greater force than the SAS patrol can counteract, so staying beyond the enemy's observation is vital. It is even more vital considering that many of the SAS's missions are reconnaissance and surveillance only. Discovery by

the enemy during one of these missions essentially negates the value of the intelligence gained, as the enemy will soon realize his manoeuvres have been discovered and consequently change his timetables and positions.

Covert surveillance demands the highest strength of character. It often requires the soldier to place himself in a hide overlooking enemy positions or supply routes and stay there for days, often with restricted movement, no one to talk to, facing crushing boredom, all against the backdrop that sudden discovery might plunge the soldier into a fight for his life. Unfortunately, the amazing stamina of someone remaining hidden for days on end in freezing or baking-hot weather accrues a lot less glamour than than high-visibility actions like the Princes Gate raid or the Bravo Two Zero patrol.

During CT, the DS will be observing the recruit closely to see whether he has the mental stamina to perform certain intelligence-gathering operations. Operating behind enemy lines requires a whole new way of moving, thinking and seeing, and the CT will give a good initial insight into whether a soldier has these talents.

The first requirement is for movement discipline during insertion to and withdrawal from an Observation Post (OP). It will usually be the four-man patrol that forms the basic recce unit, and they will have to obey the fundamentals of covert movement. Firstly, normal routes of travel which usually attract people – roads, bridges, footpaths – will generally need to be avoided, though tracking parallel to these through thick undergrowth can be a way of staying hidden yet making navigation easier. Before leaving one piece of cover the soldier should have a clear idea of where he is going, how long it will take him to get there, and where he will go if he is spotted. If hills need to be crossed, the unit will never pass across the summit, on which they would be silhouetted and quickly seen; rather they will traverse it three-quarters of the way up its height. Most importantly, the patrol has to be exceptionally diligent in picking up its rubbish – and that can include bodily functions. Anything which leaves a tell-tale sign of human presence can be an indicator of a covert unit in the enemy's midst, and may result in heavier patrolling of an area. This means

A .—	M — —	Y —.— —
B —...	N —.	Z — —..
C —.—.	O — — —	1 .— — — —
D —..	P .— —.	2 ..— — —
E .	Q — —.—	3 ...— —
F ..—.	R .—.	4 ....—
G — —.	S ...	5 .....
H	T —	6 —....
I ..	U ..—	7 — —...
J .— — —	V ...—	8 — — —..
K —.—	W .— —	9 — — — —.
L .—..	X —..—	0 — — — — —

ABOVE: Morse code is a dated part of SAS communications, and because of its accessibility other codes tend to be used during operational situations. However, it must still be known by trooper and signaller alike and is still used to deliver burst morse messages.

that each soldier must be very sure that he has picked up all his pack contents after making camp for the night or stopping to conduct surveillance. Anything left behind during a CT exercise will invoke the wrath of the DS.

When operating in a natural environment there are several key principles of camouflage which the trained SAS soldier should follow to ensure that he is fully obscured from detection:

SHAPE

In nature, there are no perfectly straight lines. Therefore the soldier should use camouflage material to break up any straight lines amongst his kit,

RIGHT: An SAS trooper and his M16A2/M203 assault rifle is silhouetted by the setting sun while training in the desert in Jordan. On operations, no SAS soldier would allow such an outline to become visible – it would give his position away instantly.

particularly rifle barrels and radio aerials (strips of natural-coloured cloth are one of the best ways of blurring an outline).

SILHOUETTE

The soldier should never present himself against a backdrop whereby he stands out in sharp relief.

SHINE

Soldiers tend to carry many metal objects which can reflect sunlight. A Special Forces soldier should ensure that anything which can reflect light is blacked out with polish or camouflage colouring.

SMELL

Smell tends to be overlooked when soldiers first join CT. Smokers, for example, might not smoke in the field but their clothing could be impregnated with tobacco fumes. These will become reactivated when the soldier sweats, leaving a tobacco scent upon the air which could be detected by a sentry. Other scents include aftershave, strong-smelling soap as well as the after-effects of certain foods (for example, garlic and other heavily spiced foods).

SOUND

When on surveillance ops, everything should be conducted in absolute silence. CT gives instruction in a common body of hand signals which enable the soldiers to communicate effectively in the field without verbal interchange. Furthermore, any items of equipment which might chink against something else or rattle (parts of a gun are usually the worse culprits) need to be taped in place.

SHADOW

There is no point hiding behind a good piece of cover if your shadow is clearly extending beyond it and announcing your presence. A good SAS soldier will treat his shadow almost as if there is another presence beside him which must also be obscured. On days when sunshine is particularly bright it is

LEFT: **The GPS system has created a revolution in military navigation. Using triangulation between 24 orbiting satellites, this hand-held set tells the trooper his exact coordinates to an accuracy of 16m (53ft), and is a useful tool for accurately guiding air and artillery strikes.**

sometimes best to wait to move until either the sun is directly above (when shadows will be at their smallest) or when the sun has set.

The combination of all these factors makes covert movement a mentally, as well as physically, exhausting process, and the CT training will weed out those who cannot keep up the discipline for hour after hour (though many such people will have already gone during the Selection phase).

Perhaps the time when this discipline is most needed is when the soldier has to set up and occupy a fixed Observation Post (OP). Constructing an OP is one of the essential skills taught during CT. It will basically consist of a diligently obscured hide, perfectly blended with the natural background (if the operation is in a wilderness area), preferably by using the materials natural to the area (remember that not all vegetation blends in well – for example, do not use evergreen material from another location and transport it to a deciduous area). The hide should be as flat to the ground as possible and the observation hole should be the minimum size possible. Woodlands overlooking the area to be watched can be some of the best locations for OPs, but the SAS solider is trained to make them in everything from flat desert plains to snowscapes.

Once in the hide, a system of resupply using the other patrol members needs to be established. This is vital as the occupier may have to spend days in an uncomfortable prone position in extremes of weather. Even though the boredom and discomfort may be crushing, the soldier must keep alert and train his senses to pick up on vital intelligence. He will also need to have a good visual memory to recall what he has seen under later questioning plus an efficient method of communicating back to the relevant military bodies.

COMMUNICATIONS AND NAVIGATION

All SAS soldiers during CT must attain the standards of a regular British Army signaller. This will involve a mastery of basic systems of coding and decoding, morse code, modern transmissions technologies, and transmission etiquette and technique. This is a demanding part of the course. The basic radio set is the PRC 319, a very powerful radio with functions such as the ability to store 20 pre-set channels, a 500-hours memory storage facility, and burst transmission. The latter is a vital part of SAS communciations. Messages are typed into a key pad, checked, encoded, and then compressed and transmitted in a fraction of a second. The receiver then picks up the burst and scrambles it out to its full length before decoding. Burst transmission means that messages can be sent incredibly rapidly and avoids the dangerous practice of staying on open-broadcast too long. The PRC 319 is just one of a series of radios, transmitters, beacons and tracking devices the SAS soldier will encounter during CT, and he will need to show a good technical competency with communications technology and techniques to pass through the course.

He must also be able to call in artillery and airstrikes competently, especially using the laser designation system. This is used in conjunction with attack aircraft launching laser-guided munitions (or 'smart bombs'), as was shown to such devastating effect in the Gulf War when the Special Air Service

The soldier must be able to call in artillery and airstrikes competently, especially using the laser designation system

helped Allied aircraft destroy enemy armour, missile launchers and communications centres. The SAS trooper holds a powerful laser beam on a target, this bounces off the target and is received by the attacking aircraft. The aircraft then releases its laser-guided bomb which continues to fly down the reflected laser beam directly onto the target. The precision can be stunning, and smart bombs are becoming a much more important part of air-strike tactics as indiscriminate area bombing becomes politically ill-advised.

Another system the SAS recruit will have to become acquainted with is the Global Positioning System (GPS) for accurate navigation. This will be an eye-opener to those recruits who have settled into the dependency on map and compass. The GPS system takes readings from a group of navigational satellites which are permanently circumnavigating

the earth. The GPS receiver – in the case of the SAS usually a Precise Lightweight GPS Receiver – measures the distance between itself and these satellites, and then uses the measurements to plot its exact position on the ground, up to a stunning 16m (53ft) accuracy. The GPS is not only a vital tool to prevent soldiers getting lost in the field, especially in featureless desert or snowy terrain, but is invaluable in providing the precise coordinates to launch air and artillery attacks.

PARACHUTING

Parachuting can be one of the final elements of CT, so if the recruit is still present at this stage then he will be feeling an increased sense of excitement. But he must still focus his efforts. Parachuting is not an easy skill (at least, parachuting well), and the risk of injury always hangs over those who undergo the course. Old injuries picked up during Selection may resurface, brought back to life by hitting the ground at around 32km/h (20mph) with a heavy pack.

The parachuting taught at RAF Brize Norton to SAS soldiers is static-line parachuting only. This means that the parachute is deployed automatically by means of a line connected between the rip cord and the aircraft out of which the soldier has just jumped. The virtues of static-line parachuting are that every soldier opens his parachute at the same height, thus helping accurate distribution patterns, and also that it is safer since deployment mistakes are less likely. Only when a recruit finally becomes a full member of the SAS will he be able to learn the advanced parachute techniques of HALO and HAHO (see Chapter 6).

The soldier needs to make eight jumps in total before officially passing Continuation Training. The first week at least will be spent on the ground. The training during this phase will teach the fundamentals of how to pack a parachute, how to land correctly, how to exit an aircraft without getting slammed against its fuselage, as well as some of the theory about how parachutes move through the air.

USING CLAYMORE MINES

During training in ambush technique, the recruit will be instructed in how to use the infamous Claymore mine as part of an ambush system. A Claymore directional fragmentation mine consists of a curved charge of plastic explosive into which is embedded 350 metal balls. The charge is usually mounted on a pair of spikes, these being stuck into the ground with the convex face arched outwards towards the killing ground. When detonated, the metal balls will spray out at high velocities over a 60-degree arc at a height of around 2m (6ft), and with a killing range of around 50m (160ft). Any enemy personnel within this arc have little chance of survival, especially as the SAS are taught to detonate several Claymores in tandem to cover all possible avenues of escape. Used extensively in Vietnam, the results of a Claymore mine being detonated are invariably horrible: the victims are almost shredded by the blasted metal balls. Yet they are a very useful ambush tool, primarly because the arc of the bomb shape produces a wide angle of attrition which it would take many seconds for a machine gun to sweep adequately.

Furthermore, the Claymore also has an anti-vehicle contribution, as it is able to shatter the engines, tyres and equipment of light vehicles and trucks. An SAS unit will detonate the Claymores while simultaneously opening up with heavy small arms fire, the effect being a truly lethal killing zone.

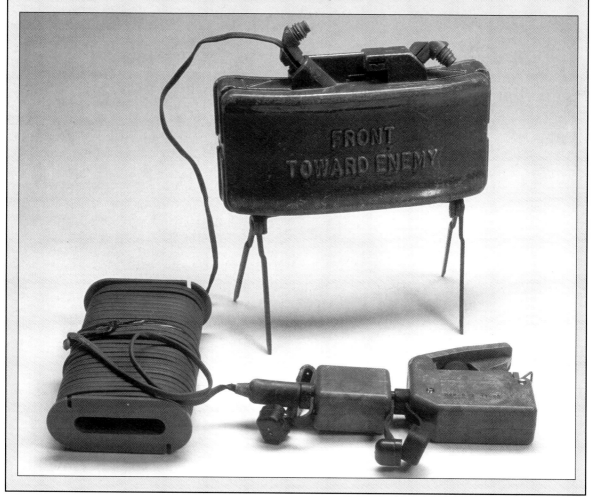

SAS soldiers experience fear like anyone else. However, their motivation to complete the jumps is stronger than that of other people

Recruits are also taught about the correct drill aboard an aircraft and immediately after making the jump (such as looking up to make sure that the canopy is not twisted). The most time will be devoted to the art of landing, the most dangerous part of the drop. An SAS soldier dropped behind enemy lines who has just broken his ankle is likely to end up dead or captured, so the instructors (actually RAF Parachute Jump Instructors) make sure that the soldiers can manage rough landings coming in from any direction.

The second week will shift its emphasis to exit training on aircraft mock-ups at ground level. This will include jumping from the mock-ups while being buffeted by the wind from enormous fans which simulate the slipstream of a real jump. The procedure aboard the mock aircraft is repeated exactly as it will be in the air. Further exit training takes place from a 223m (75ft) tower. The recruits are restrained by harnesses which give them the feel for what it is actually like to experience the impact of a parachute opening.

By the end of the second week, the recruits will have been introduced to the various items of parachute kit and the correct way to wear their equipment during a jump. The kit includes the standard Irwin PX1 Mk 4 parachute with PR7 reserve and also a custom-designed life jacket to be worn during the water jump. The bergen is worn underneath the reserve chute, but is released on a line to hang under the paratrooper once he is in the air. This allows the bergen to hit the floor before the parachute jumper, thus significantly reducing the impact on his legs and body.

The first jumps take place in the third week. The recruits will have to jump in two columns of men from a C-130 Hercules flying at around 300m (1000ft). The procedure follows the same format:

● The order to check kit is given. The recruit checks his own kit and then the kit of the man in front of him.

● The recruits are ordered to stand and they advance in two lines to face the doors at the rear of the aircraft.

● Each man hooks up his static line to the cable running along the inside of the aircraft.

● The light over each door changes from red to green and at the instruction each recruit leaps hard through the door and commences his drop.

If this jump is completed successfully, with no injury on landing, then the recruit can face the next seven jumps with a little more equanimity. Yet the learning curve is steep. The altitude of the drops will become radically reduced, in cases down to approximately 150m (492ft). Another drop is made into water, the soldier having to extricate himself from his tangled lines and parachute as he bobs up and down in the sea. The final jump takes place at night, a scary experience according to one SAS soldier:

'Even in the Regiment blokes are terrified of parachuting at night, especially the biggest guys. They come down heavier, you see. If they drop you in the wrong place, over power lines and water, you can drown or break your neck. Having said that, at the end of the day it's the coward's option to jump. Does that sound surprising? You have to remember that you are frightened of what people, your mates, are thinking about you. At the end of the day it's easier to jump, despite the risks.'

As we can see from the above account, SAS soldiers experience fear just as anyone else does. However, their own motivation to complete these jumps is stronger than that of most other people. In spite of recruits' fears, fatalities are extremely rare during the parachute course – the SAS has had only one.

The added motivation to complete the night jump is that if you manage to do it successfully, and you have no other problems with your course performance, then you can be 'badged', a member of the Special Air Service. We will say more about this transformation in the final chapter, but now we must look at the other side to CT – combat survival training.

Combat Survival Training

Combat Survival Training is designed to give SAS soldiers total self-sufficiency, regardless of the terrain. From the snow-laden hills of South Wales to the sweltering jungles of Brunei, CST teaches the SAS recruit to live off the land and stay alive without re-supply indefinitely.

Much of the content of Continuation Training is instructional and theoretical, and does not approach the physically demanding practices of Selection. However, there is one aspect of CT that, sometimes controversially, places the SAS recruit under extreme physical pressure, as well as an equivalent mental load. This is called Combat Survival Training (CST).

CST is designed with several intentions. First, it teaches the recruits the basics of how to live off the land if they have to. SAS soldiers operate far from logistical resources, and it is all too easy for them to run out of food and water if a mission extends beyond its natural limits or goes badly wrong. Left exposed to the elements, and without sustenance, they must be able to stay alive and healthy through

LEFT: A jungle shelter constructed during jungle-warfare training. Note the positioning of the fire, far enough away from the shelter to avoid significant smoke inhalation and the risk of fire, but close enough to act as an effective repellent to flying insects.

131

Special Forces soldiers are extremely desirable personnel for the enemy to capture... they will know that the soldiers are privy to genuinely useful military information

survival means alone. Second, the Escape and Evasion part of the course focuses on how to escape captivity, evade captivity, and also resist interrogation within captivity. All these elements are vital, as Special Forces soldiers are extremely desirable personnel for the enemy to capture, and because they operate so far from conventional forces they are also unlikely to be able to successfully pass themselves off as anything else. Furthermore, the enemy will know that the soldiers are privy to genuinely useful military information. They will often try to extract this type of information through violent means, such as torture, or through psychological means. Interrogation training teaches some skills for resisting and evading the interrogator's techniques. Perhaps more importantly, it also enables the staff of the Joint Services Interrogation Unit (JSIU) to assess whether the soldier has the mental toughness to handle some of the worse excesses of interrogation. CST essentially separates into three sections with three locations. The first part is held at Hereford, and introduces the recruits to the fundamentals of survival. They are instructed in the basic principles of finding food and water, making shelter and fire, and also how to accomplish all this while still remaining undiscovered by the enemy. The second part is the Jungle Training. Jungle Training is run twice a year (usually beginning in March and September) by the SAS, and it takes place at the British Army Jungle Training School in Brunei. In total the course lasts about six

RIGHT: Here we see an SAS trooper during the Falklands War in 1982, dressed to combat the sub-zero climate. He wears a smock and trousers in Disruptive Pattern Material (DPM) with an SAS combat and survival belt kit, and a balaclava helmet for head protection.

weeks, giving the recruits a good idea of how to live and fight within the densest jungle terrain.

The third part of CST is conducted back in the UK at Exmoor and other locations across the UK. This is the controversial Escape and Evasion course. The recruits are taught how to evade capture far behind enemy lines. However, the part of Escape and Evasion which raises most eyebrows is the Resistance to Interrogation training. This is conducted using personnel from the Joint Services Interrogation Unit, which specializes in putting prone-to-capture servicemen and women through some extremely realistic, even brutal, interrogation scenarios. If at any stage the recruit looks as if his resistance might crumble and he might give away essential information, then he will be RTU'd just as if he was in the first week of Selection.

These three different elements of CST might not appear in this same order with each course, but taken together they will give the SAS recruits a thorough grounding in what it is necessary to survive in some of the world's most hostile military and natural environments.

SURVIVAL TRAINING

Because the Jungle Survival course is a distinct section of Combat Survival Training, it is worth dealing with separately. Firstly, we shall look at the survival skills which the recruits will be taught in the UK. Survival training is a long process within the SAS. The demands of terrains such as deserts and polar landscapes are quite different to those which would face a soldier in a temperate climate like that of the UK. Thus even after the recruit has passed CST to become a fully fledged SAS soldier, he will still receive survival training relevant to his possible future areas of operation.

However, the UK survival training does cover the fundamental principles of survival which can be taken to most areas of the world. These principles are as follows:

FOOD

The recruit must be able to live off the land from both plant and animal sources. In reality, this means knowing how to identify edible and poisonous plants, how to catch, kill and prepare animals for cooking, general methods of survival cooking, and different ways of storing food.

WATER

Though humans can survive in the right circumstances for over a week without food, if they are deprived of water then they will be lucky to survive even a few days. Finding water is consequently one of the most important skills taught to the recruits on CST. They must be able to find natural sources of water, extract ground- or air-held water through artificial means, understand the principles of balancing food and water intake, and know which plants contain water.

SHELTER

Whatever the climate, shelter is vital. Shelter can be

FINDING SOURCES OF WATER

The following are good indicators, even in desert regions, of where water might be found:

● Swarming insects, especially bees and ants, may often indicate that a water source is nearby (either pooled or underground).
● Fresh vegetation growing in an otherwise dry landscape can mean that there is water beneath the surface. This effect is often seen on dry riverbeds, particularly at the corners of bends, where the surface water is usually the last to recede.
● Birds tend to drink water at dawn and dusk. If a bird has already fed it tends to fly shorter distances before resting on a tree or another perch. See if you can work out where it has come from or where birds are going to, but be careful as birds can fly long distances to water.
● Like birds, grazing animals tend to travel to water at dawn and dusk. This sign can be useful in African regions where there are clear patterns of herd movement. Alternatively, look for clear patterns of animal tracks.
● Rocks with a damp surface may indicate that there is a spring beneath them. This is particularly the case with limestone or lava rocks.
● On coastlines, try digging above the high water mark – fresh water is often filtered out by the sand and floats on top of the saltwater.

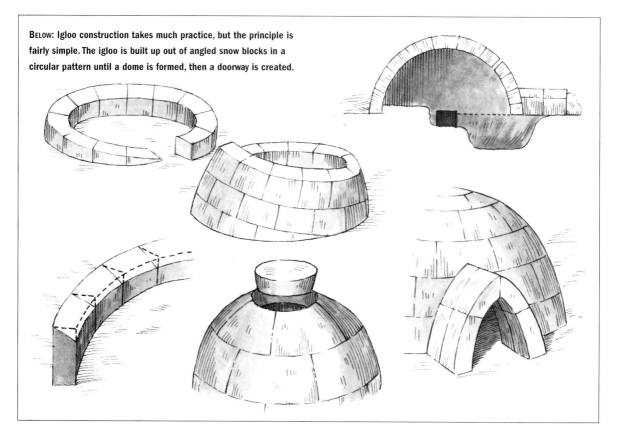

BELOW: Igloo construction takes much practice, but the principle is fairly simple. The igloo is built up out of angled snow blocks in a circular pattern until a dome is formed, then a doorway is created.

anything from the canopy of a tree to a sophisticated branch and twine construction, but it must protect the occupier and his equipment from the elements. During Combat Survival Training, the recruit is instructed how to build a basic shelter from natural materials, choose an appropriate location for his shelter, adapt a shelter to wet and dry conditions, and make his shelter habitable and suitable for cooking in.

FIRE

Fire is a precious element of survival. It enables cooking, it gives direct heat and light, in hot climates the smoke can help to keep away insects, and it can also boost morale. However, elite soldiers must be cautious about building fires because they are an ideal locator for the enemy. On CST the recruit is told where and when not to build a fire and, most importantly, how to start a fire without matches or a lighter.

Taking each of these elements in turn, we shall look at some of the most important techniques for a recruit to absorb during CST. At all times the recruit must remember that later during the survival course he will be left in the countryside, with only the clothes he stands in, to survive for several days. Beforehand, his clothes and body are rigorously searched for anything which might help him survive – such as a penknife or a box of matches – and these are routinely confiscated. Nothing must give him an unfair advantage. His ability to survive his period of isolation is directly related to how well he has learnt the skills over the previous weeks. If he has to be rescued, or copes badly, he will be RTU'd.

RIGHT: In all survival environments, finding clean drinking water is the paramount priority. This soldier demonstrates a good method of catching rainwater, using a structure of twigs to support impermeable leaves, these forming a 'cup' in which rainwater collects.

FINDING WATER

Water is the most essential element for survival. If the soldier on survival exercise – or in combat – cannot make his water intake greater than his water output through sweating, digestion and respiration, then his health will quickly deteriorate. Even if you were to sit still for the entire day, doing no form of exertion whatsoever, your body would still lose about one litre (nearly two pints) of fluid, and that rate jumps to two to three litres (three to five pints) during normal levels of exertion. Take a soldier in a blistering hot desert marching miles through the heat with heavy packs, and you start to have some idea how quickly even the toughest soldier could dehydrate. Add some form of fluid-demanding illness – such as diarrhoea or vomiting – and severe dehydration could result in hours with death shortly following.

Accordingly, one of the first principles you will be taught is how to conserve the water you already have within your body. Some pieces of advice are fairly obvious. Rest as much as possible to reduce fluid loss through sweating. Stay out of the sun as much as possible and try to do your travelling at night. Keep away from very hot surfaces such as rocks exposed to the sun. If you do have a water supply, ration it for the next couple of days and be very disciplined in this respect.

However, some facts about water conservation are less well known. You should restrict your talking as much as possible, as the respiration involved increases the fluid loss when you breathe out. If your water supply is very low, do not eat. Digestion requires a large amount of fluid to complete its processes, particularly fatty foods. You could literally eat yourself to dehydration. Only eat if your water supply is there to match – a reasonable guide is if you can afford the water to wash your meal down substantially and drink throughout the day then you should be OK to eat. Finally, do not smoke or drink

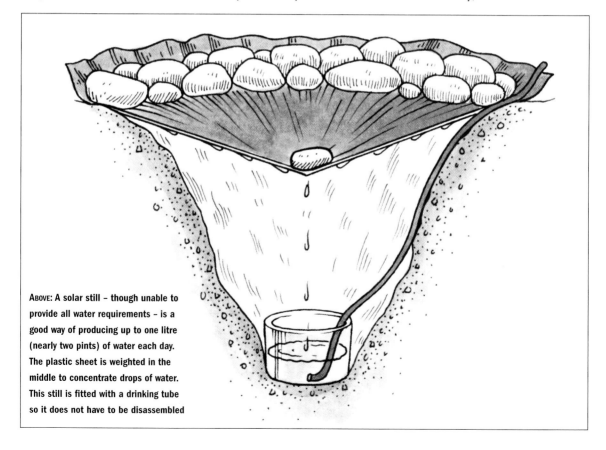

ABOVE: A solar still – though unable to provide all water requirements – is a good way of producing up to one litre (nearly two pints) of water each day. The plastic sheet is weighted in the middle to concentrate drops of water. This still is fitted with a drinking tube so it does not have to be disassembled

RIGHT: During the Gulf War, SAS clothing had to be adapted for blistering hot days and chilly nights. This soldier wears a camouflaged jumper and combat trousers. He has fingerless gloves for extra warmth. The camouflaged mesh net is useful for keeping off insects.

alcohol, as both will accelerate fluid loss (an SAS soldier is unlikely to carry alcohol on operations, but smoking is more common).

In temperate zone countries, such as the UK, finding supplies of natural water is rarely difficult. Apart from the plentiful man-made sources (a covert mission to find a farmhouse water tap can be useful), temperate zones usually have streams, rivers and rainfall in ready supply. However, this is not always the case, and a recruit is trained to find and extract water under even the most barren conditions.

If an open water source is found, extreme caution must be exercised even if your thirst is raging. Many water sources can be poisoned by either natural or man-made toxins. Signs of poisoning can include: a foul smell from the water and noxious bubbles on the surface; dead animals and no fresh plant life around the water source; lakes which have no fresh water inlet or outlet. Whatever the appearance of the water you should purify it before drinking. Ideally this should be done with purification tablets, which should be an essential part of you survival kit. However, if these are not available then filtering the water through a piece of cloth prior to boiling it for ten minutes should remove most impurities.

Of course, sometimes there is not the time for this sort of preparation. Tentative sips can be taken from a free-flowing river which runs clear over a limestone or shingle bed, but an even better ready-made source is rain water. Collect rain water whenever it falls and save it if possible. Soaking a piece of clothing and then wringing it into a container is one of the best methods. Alternatively, look for natural reservoirs such as hollowed-out trees or cup-shaped plants which may contain rainwater. Plants themselves – for example, bamboo, cactus and palms – might also be water-storing. The soldier should study the types of plant that thrive in his area of operation before he arrives.

Sometimes the soldier will find no openly accessible sources of water. This leaves three options:

extracting water from underground, extracting water from condensation, and finding plants which hold their own water supplies. Finding water underground is not easy, and takes a trained eye to recognize the signs that water is in fact present. The best places to look are dry river or stream beds – though the surface water may well have disappeared, water can remain underground for fairly long periods of time. Choose the lowest point of a bend in a stream or river bed and dig down – after a few feet you might find water starts to fill up the hole.

Bushmen in the Kalahari Desert in Africa use a technique of digging down until the ground goes moist, and then inserting a reed into the ground and sucking on it for up to ten minutes. Eventually they find that water is drawn through the ground and up into the reed or straw.

Condensation is a much more complex way of obtaining water, but it can produce life-saving results. There are two basic methods of this form of water collection: the solar still and the transpiration bag. Both require the use of one or two sheets of clear polythene (such as from a strong shopping bag). The solar still has a valuable application in desert regions, whereas the transpiration method requires you to find fresh green vegetation.

To make a solar still, first dig a hole about 90cm (36in) across by 45cm (18in) deep. Put some form of receptacle at the bottom of the hole to catch water, then place the plastic sheet over the hole, weighting it down around the edges with stones or sand to keep it in place. The sheet must dip down in the centre, without touching the earth sides of the hole, so that the lowest point of the dip is directly over the receptacle. Placing a single stone in the middle of the sheet as a weight will achieve this. The solar still works by trapping the condensation between the ground and the sheet as the sun warms up the ground throughout the day. The water vapour condenses on the underside of the plastic sheet and then runs down to the lowest point (defined by the single stone on top) then drips into the container. Properly constructed – remember to seal the edges of the sheet tightly so that water does not evaporate to the outside – the solar still can produce around one litre of water each day. This is not enough to keep you alive, but

combined with other methods (or using several solar stills) it could be a vital part of your water supply.

A transpiration bag works on much the same principle, except in this case it extracts water vapour from plants instead of the ground. Tie the plastic bag around a section of bushy, green and healthy vegetation, taking care that the sides of the bag are not pressed against the leaves. Once again, when the sun warms the bag and its contents up, the water vapour will collect against the sides of the bag and run down to the lowest point, usually the corner. You then collect the water for drinking. The same technique can be used on vegetation which has been severed from the source tree or plant. Place the vegetation in the bag, but sit it on a bed of stones to stop the vegetation from touching the bag floor. Stop the 'roof' of the bag doing the same thing by propping it up with a twig which has

Some plants will provide a highly nutritious and filling meal, while eating others will result in a slow and painful death from poisoning

a stone (stone will not absorb the water) between the tip of the stick and the bag.

Using these methods, the soldier should be able to supply most, if not all, of his needs for fluid intake. Once this intake is sufficient, then he can turn his attention to food.

FINDING FOOD

Finding food in a survival situation is an enormous topic. Almost every country in the world has plant and animal species, or combinations of those species, unique to itself, and thus the natural larder shifts wherever you go. SAS recruits on CST will learn how to acquire and prepare commonplace sources of food in agricultural regions – pigs, sheep, cows, and arable crops. However, these food sources are often accompanied by a human

RIGHT: This trooper is wearing arctic warfare gear. It consists of various thermal layers over which is worn a waterproof and windproof hooded jacket, and trousers. The bergen also comes with a white arctic cover. The mittens are specially adapted for normal use of the firearm.

BELOW (B): This trap uses a simple counter-weight to hold the snare in place. When the animal puts its head through the snare and is trapped, the counterweight will either lift the animal off the ground (if it is light enough) or hold it in place through the cross-member jamming in the V of the tree.

ABOVE (A): A snare trap using a bent sapling to hoist the prey off the ground. The sapling is bent over and anchored using a hooked stick slotted into a notched stick stuck in the ground. A snare loop is placed over an animal track. When triggered, the hooked stick pulls out and the sapling jerks upright with the prey.

presence, making them an inadvisable food source for soldiers looking to be invisible to the enemy.

The CST course will help the recruits gain some understanding of how to find and prepare natural food stuffs. The recruits are usually taken to see a professional botanist or agriculturist who will show them around a garden stocked with the type of plants they might eat in the wild. They will also get the chance to kill and prepare some of the animals most readily accessible to the survivor. It is not possible for us to list all these plants and animals, but we can explore the basic teachings of the course.

PLANTS

Plants are by far the most easily gathered food source for the surviving SAS soldier. Yet the very fact that the earth is populated by literally millions of different plant species makes it harder to identify which are edible and which are not. Some plants will provide a highly nutritious and filling meal, while eating others will result in a slow and painful death from poisoning.

What is required is a simple test to check for the plant's edibility. The SAS has the Universal Edibility Test (UET) – and it is one followed by Special Forces soldiers of other countries throughout the world. But prior to using the UET, visual clues and general species principles alone can give you a reasonable idea of whether the plant is edible or not. These principles are:

● Do not touch plants with white or yellow berries as they are generally poisonous. Plants with blue or black berries are almost always safe to eat and about 50 per cent of red-berried plants are safe to eat.

C

BELOW (D): If an animal lives within the trunk of a tree, carefully position the snare around the entrance and secure it in place with a loop around the trunk. Make sure that when positioning the snare you do not disturb too much of the surroundings, as animals are very sensitive to changes in their home environment.

D

ABOVE (C): The snare is tied to the upper end of a branch and this branch balanced on a notched stick so that the snare hangs down over an animal track. When triggered, the branch comes off the stick, restraining the animal through weight, or through a deadfall action.

● Do not eat a plant with a milky white sap.
● Steer clear of wild beans and peas. These have a tendency to withdraw minerals from the soil which in turn causes stomach upsets.
● Do not eat any plant with umbrella-shaped flowers, except carrots, parsley and celery. (Note: celery should be an addition to a diet as it takes more calories to digest than it provides.)
● Do not eat any plant which irritates your skin.
● Single fruits on a stem are usually safe to eat.
● Do not eat old plants or any which appear to have been severely attacked by insects.

The soldier should be able to make an informed decision about whether to eat something or not. However, performing the UET gives a much more sophisticated level of assurance. It essentially works by gradually introducing the body to the substance and monitoring it at each stage for any adverse reaction. The test follows these progressive stages:

● Take one part of the plant only (don't mix substances during the period of this test as you will not know which part/plant caused any reaction) and, on an empty stomach, rub it on your wrist. Wait about 15 minutes to see if there is a reaction.
● Put a small part of the plant on your outer lip. Wait for a reaction.
● Put the piece of plant inside your mouth, on your tongue, and hold it there for about 15 minutes. If there is no reaction chew it thoroughly, keep holding it in your mouth and wait again for a reaction.
● If you have had no adverse reaction, swallow the substance and wait for several hours (preferably about eight) to see if you suffer any signs of

A figure four trigger, effective and easy to make

Release trigger

A deadfall with a tripline release trigger

A deadfall/snare combination

A trigger for a deadfall/snare combination

illness. (If there are make yourself vomit and drink lots of water.)

● If you still feel fine, eat a small handful or cupful of the plant and wait another period of several hours to see if there is a reaction. If there is no reaction, you can class the plant as safe to eat.

The UET and the principles of edible plant identification should enable the recruit to have a solid approach to identifying and consuming plantstuffs. (The one exception is fungi – the UET cannot be safely applied to any type of fungus.) Cooking plants will essentially consist of boiling or roasting over an open fire, but be cautious because plants can change their chemical properties during cooking and consequently become poisonous.

ANIMAL FOOD SOURCES

Animal food sources almost always need cooking to be safe to eat. In a country such as the UK, food types are mainly rabbits, pigeons, deer, farm animals (particularly sheep) and pigs. In Africa the number of edible animals increases tremendously. However, the survivor – effectively an intruder into the animal's world – should always exercise caution. A buffalo would provide food for many days (even weeks if the meat was properly dried and stored), but unless you were able to shoot it at a distance the options for tackling such a powerful and aggressive beast would be limited.

Ideally, the survivor should aim to kill most of his food by either remote means – such as traps – or by projectile means which allow some distance between the animal and the hunter. Tackling animals up close can be dangerous, and one thing the survivor must avoid in the wild is an injury. Unless the hunter is sure of what he is doing, even the docile sheep can cause injury to legs and arms in the tussle to sink a knife into it (though this is easier in flocks which come into regular contact with human beings and are used to being handled).

Recruits to CST will be instructed in a variety of methods of survival hunting and fishing, and also the requisite skills for cooking the creatures. Here we can look at the basic principles of survival hunting as regards land mammals and some general methods of cooking.

Animal traps essentially separate themselves out into four categories: snare traps, deadfall traps, spear traps and net traps. Snare traps consist of a wire or strong noose which is tied in a loop with a slip knot. The aim is to make the animal – such as a rabbit – put its head through the noose at which point it tightens the noose around its own neck and strangles, or at least restrains it, until you return to the trap later. Snare traps can be of any level of sophistication, but the most basic is a simple loop just slightly larger than the head of the animal you intend to catch, and positioned along an animal run or at the entrance to a burrow. Make sure that the snare is securely anchored, as the animal will struggle to get free, and even a rabbit can be quite strong. The humble snare can be made even more effective by linking it to a sapling or branch under tension; when the snare tightens around the animal's head and the animal pulls, it dislodges a restraining device on the branch. The branch is then released under tension and hoists the creature into

ESSENTIALS OF SURVIVAL FISHING

Fish can form one of your most accessible and nutritious sources of food in a survival situation. Here are some SAS tips for effective survival fishing:

● Make hooks out of any sharp objects available, such as pins and belt buckles.

● Try to find bait that is similar to the diet of the fish in their natural habitat. This can include worms, maggots, and flying insects.

● Make a simple fish trap by cutting the head off a plastic bottle and then inverting the head inside the bottle. The fish can then swim into the bottle but swimming out is much more difficult.

● Narrow a stream with stones or mud to channel the fish into a net trap.

● When attempting to spear fish, have the tip of the spear already in the water so that the creature is not startled by an explosive splash.

● Improvise a fishing rod out of a piece of branch and line and use it for fly fishing. Float a large, wriggling insect on your hook to attract striking fish.

BLOOD CHITS AND BLOOD MONEY

When SAS soldiers embark on operations into enemy territory, they are given two items which may ensure that any civilians they meet on the way (and even some enemy soldiers) are willing to help them pass through safely.

The first of these is known as the Blood Chit. This is a promissory note which – in relevant languages – tells the recipient that if he aids the soldier in an escape, evasion or mission, then he can present the note to a British Embassy or consulate and claim a substantial cash reward.

A more immediate incentive is in the form of Blood Money. Blood Money is literally a monetary bribe, usually in the form of gold sovereigns (high value for low weight carried) but also US dollars, and is an essential compliment to the Blood Chit as, particularly in developing countries, many cannot read. Thus in the Gulf War SAS troops were issued with 20 gold half sovereigns each.

the air where it will be less able to struggle. The restraining device can be a small piece of wood, tied by string to the branch in one direction and the snare in the other, held in place by carving a notch in its surface which hangs against another notch in an anchored piece of wood on the ground.

Deadfall traps are basically powerful crushing tools. Again a trigger system is used, but the deadfall trap simply deploys a heavy weight which drops onto the animal and stuns or kills it. Deadfall traps take some time to perfect, but the most basic version consists of a log, propped up by a branch. The branch in turn is connected to another branch or a wire which is attached to a piece of bait beneath the log. The theory is that the animal takes the bait to feed on, thereby disrupting the supporting branch, which in turn leads to the main log dropping on it and killing it.

CST will introduce the recruit to some of the more basic deadfall traps. The virtue of the deadfall is they can be scaled up to take on some substantial prey, including wild pigs and deer (they were also used against humans during the Vietnam War).

Spear traps can also be used against a variety of sizes of prey. As their name implies, spear traps literally spear the animal. The essence of the trap is a springy sapling or branch to which is tied one or many sharpened wooden spikes. The branch is bent under tension and then tied in place by a tripwire. The tripwire falls across an animal track. When the animal hits the tripwire (it must be disguised to be successful) the branch is released and the animal is hit by the spikes. Spear traps are useful in that the animal is generally either killed or not - there is little chance of it escaping after being hit. On the down side, the animal has to be in exactly the right place for the spikes to impact upon it.

The final type of trap is simple netting. On land, this is primarily used for catching birds. Netting is stretched between trees on a birds flight path, and the birds simply fly in and get caught. If netting is not available, then just small pieces of line stretched between trees may suffice to injure the birds and leave them for collection from the woodland floor.

With all traps, they have to be convincingly and competently made to have a realistic chance of catching food. All traps must blend in with the background using natural and local camouflage otherwise the animal will know that something is wrong. When making the trap the materials should be touched as little as possible (or ideally wear gloves) as your scent will be detected by the animal and it will avoid the trap area. Sometimes, it is simpler for the survivor to make projectile weapons. A piece of elastic and a Y-section of a branch will make a fine catapult which, with practice, can be used to kill everything from rabbits to fish. Spears can be constructed out of long branches with an arrowhead made from a piece of chipped and shaped stone and flights from leaves or bird feathers. A bow and arrow is a more complex piece of engineering, but the principles are readily apparent. Whatever the tools used, the survivor - and the recruit during CST - must weigh up the advantages of making one over the other. Remember, the purpose is not to

RIGHT: A soldier prepares to boil up a mess tin of water during survival training. He is using fuel tablets, these having the advantage over wood fires in that they produce minimal smoke - which might alert an enemy to the soldier's presence - and they are safely contained.

make a fine trap or weapon, but to catch prey and provide food.

Towards the end of CST the Regiment often holds a 'banquet' for all members of the course. This involves eating the varieties of creatures you will have learnt to capture and kill during the course – rats, hedgehogs, worms, and crickets, as well as assorted small birds. A fitting end to the course!

MAKING FIRE

Making fire without matches is one of the primary lessons of CST. All outdoor fires consist of three elements: tinder, kindling and fuel. Tinder is material which combusts extremely easily – such as dried grass, cotton fluff, and small wood shavings, essentially any light, bone dry and small material. This is used to start the fire. Kindling consists of slightly

bigger pieces of material used to let the fire take hold – small, dry twigs are the most readily available. Fuel is what you put on the fire once it is going to keep it going and bring it to its required temperature. SAS soldiers will often have survival matches as part of their kit (matches which are waterproofed with wax), but CST teaches them to make fire without such tools. Firstly, create heat or sparks sufficient to set the tinder on fire. Once the tinder is burning then add small pieces of kindling until the fire takes hold. Finally, add fuel on the fire to bring it to full strength and utility.

The main methods of fire-making taught are:

FLINT AND STEEL

Striking a flint with a steel can produce a shower of sparks strong enough to ignite tinder. Hold the flint

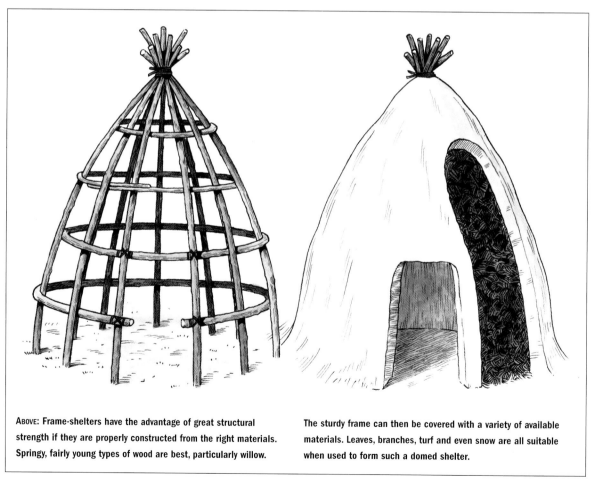

ABOVE: **Frame-shelters have the advantage of great structural strength if they are properly constructed from the right materials. Springy, fairly young types of wood are best, particularly willow.**

The sturdy frame can then be covered with a variety of available materials. Leaves, branches, turf and even snow are all suitable when used to form such a domed shelter.

close to the tinder and strike it with the steel. As soon as the tinder starts to smoulder, gently blow on it until more substantial flames appear, then add kindling. For soldiers who still have ammunition, emptying gunpowder onto the tinder can speed up the process dramatically.

HAND DRILL AND FIRE PLOUGH

These classic techniques of making fire take some practice to perfect. For a hand drill, take a piece of flat hardwood and cut a V-shaped notch into one end. Place tinder in the notch, then make a small depression in the surface of the wood just at the edge of the V. Select a softwood stick and sharpen it. Put the sharpened end into the depression and then roll the stick fast between your hands while applying downward pressure. The heat that builds up through friction will eventually produce smouldering bits of wood around the tip of the stick which drop into the tinder. Blow gently to encourage the tinder to catch fire and then add kindling. Make sure that there is a small air gap between the baseboard and the floor, as air is needed for the tinder to catch light.

The fire plough works on the same principles as the hand drill, the difference being that a long groove is carved the length of the board terminating in the V. A stick is rubbed along this track rather than spun in a hole, and the smouldering wood is pushed from the track onto the tinder at the end.

LENS METHOD

As most schoolchildren are aware, concentrating the light of the sun through a magnifying glass produces powerful heat. Simply direct the rays onto tinder to gain combustion. A telescopic sight or binoculars can be used for the same purpose.

SHELTER

The final strand of the recruit's training during CST is the technique of making shelters. We have already seen examples of how to build basic survival shelters in Chapter 3, but during CST the recruit will learn how to make far more advanced constructions. Generally, the principles remain the same. Most survival shelters begin with a natural feature, such as a rock, tree, depression, or gully, and this is

enhanced using branches, leaves or man-made materials such as plastic sheeting to make it as wind- and water-resistant as possible. Branches tied together in a lattice pattern using either twine or strong plant material form an open panelling which can then act as the basis for walls or a roof (the roof should be angled and laid with tightly packed leaves, rushes or grasses to form a protective layer). Building a two-layer roof can also be beneficial – leaving an air gap between the two layers creates a section of still air which keeps out the heat in summer and keeps in the warmth in winter, in the same way as double-glazing works.

Whatever the style of construction, as a recruit you must show the DS that you can build something that is strong and fulfils all the criteria for a shelter: refuge from the elements, storage for equipment and protection from any wild creatures. He will also be checking that you position the shelter properly. The best shelter in the world will be undone by poor location. Bad sites to place your shelter would be:

- Very exposed places, such as hilltops and flat, open fields.
- Close by a river or stream – on a hillside position directly beneath running water – as sudden downpours might lead to flash flooding.
- Underneath dead wood or loose rocks which might fall on you.
- Along tracks used by large animals.
- The bottom of valleys as these tend to be damp.
- Near insect nests.

Combat Survival Training will rarely instruct the recruit in how to make semi-permanent survival shelters such as log cabins. CST is designed instead to teach the recruit about what they would need to do if an operation went amiss and they were called upon to survive. In these difficult circumstances the shelter is a temporary stop until he is able to move or be rescued.

JUNGLE SURVIVAL

Jungle Survival is one of the most vivid and memorable elements of the SAS Selection process. The recruit suddenly finds himself transported from the dour expanses of the UK wilderness to the colour-ful, hot and frequently bizarre environment of trop-ical Brunei. Both animal and plant life have an unbelievable diversity, and there is the permanent threat from the many different poisonous species that live within the jungle.

Jungle Training lasts four to six weeks and is designed both to teach the recruit the survival skills specific to that environment and to see whether he is capable of applying them. The jungle also pre-sents a peculiar set of problems. The climate is extremely humid, so clothes and kit are permanent-ly drenched and certain pieces of equipment are under the risk of corrosion. Insect life can be painful, with mosquitoes only one of legions of insects which attack and bite. The jungle depths often make navigation near impossible, as well as having a claustrophobic and disorientating effect. The constant need to hack your way through the foliage soon means that your direction of travel wobbles constantly, while the dense jungle canopy ensures that distant landmarks and celestial bodies are almost impossible to spot. On top of this, the soldiers will have to show that they can continue to perform the duties of soldiering despite the terrain, keeping together coherent patrols, setting ambush-es, and tracking the enemy.

During jungle training, the first task is to adjust to it mentally. While the jungle can be undeniably hos-tile, it is also relatively friendly to survival in many ways. Water and food sources are plentiful, as are the natural materials to make shelters. It is easy enough to hide in a jungle environment, and ambushes tend to be extremely effective when staged properly. If you accept these simple facts and treat the jungle more as a friend than a foe it may go a very long way to overcoming the sensations of oppression which some soldiers have reported. Furthermore, endeavour to train yourself not to become frustrated when you encounter obstacles every second of your journey through the jungle. As with the exacting trials of the Selection process, just accept them and try to deal with them one at a time.

The next part of your successful completion of jungle training involves adapting your basic survival training to meet the distinctive tropical conditions. The jungle has several unique challenges:

LEFT: An SAS patrol in Sarawak carries out checks on equipment just prior to heading out on a jungle patrol during the conflict in Borneo in 1963-1966. Borneo gave the SAS the opportunity to perfect deep-insertion raids and ambushes into enemy territories within the jungle.

FLORA AND FAUNA

The jungle has plentiful supplies of animal and plant life which are edible, so food is never a problem. Unfortunately, poisonous plants and dangerous animals are in much wider abundance. Follow the instruction carefully when learning what is poisonous and what is not, and which animals to avoid. When exploring amongst trees or turning over stones do so with sticks rather than your hand in case there are hidden snakes, spiders or scorpions. Shake out your boots after you have taken them off in case insects have made their home there. Studying a book on tropical wildlife is advisable before going out to Brunei – particularly concentrate on identifying poisonous varieties of snake. If you come across such a snake leave it well alone; if that is impossible then kill it with a sharp strike to the head with a stick or your gun butt. When eating tropical plants, be especially sedulous in following the UET principles, as some tropical plants are very poisonous and can cause serious illness.

One animal you will encounter a lot is the leech. These blood-sucking creatures slide off wet leaves onto exposed skin and attach themselves to human flesh very firmly. Trying to pull them off will only result in leaving their mouth parts embedded in your skin to become infected, so instead touch the leech with a heat source – such as a lit cigarette – they will shrivel and drop off.

SHELTER

Build your jungle shelters on platforms, not on the ground. Make a platform by tying a lattice mat of branches either between four poles or, more robustly, between four trees. Platform shelters will keep the worst of the land-living insects away, particularly if you wrap the base of each support in petrol-soaked rags (though beware of fire). Hammocks are your best sleeping method for much the same reason. If not, make a bed from poles or bamboo covered in a layer of thick, non-irritating leaves. Use a mosquito net, otherwise you will have a harrowing night and an itchy day. A camp fire can also be one of your best methods of warding off insects. Build a fire, and put some damp wood on it before you bed down – the resulting smoke will keep many insects away, but be careful that your shelter has a chimney so that you are not overcome by smoke inhalation.

RIVER CROSSINGS

River crossings are a regular part of jungle and general CST training, and can be quite dangerous. (More SAS soldiers have been killed during water-related training than any other element of Selection or CT.) Tropical rivers can be fast flowing and turbulent, and even strong swimmers can be swept away – good teamwork is required. If the river has a strong current, make sure that you are secured by rope to the team on the bank before crossing. Take a long stick with you – it can be used to probe the riverbed for obstacles and also to lean against into the current if it threatens to take you. When planning your crossing route, make a diagonal course against the downstream flow of the water to your intended destination. This is because the effect of the current will move you in a downstream direction with every step. Once you are across, secure the rope to something solid and then the other members of the team can use this as a fixed guide for their own crossing. When everyone is across check each other thoroughly for leeches and other parasites.

Another method of river crossing or travel is a raft. It is fairly easy to build one using the plentiful supply of logs found in the jungle bound together with rope or vines. Remember to cut notches into the wood for the ropes to sit in, otherwise they might start slipping once you get into the water and the raft will disintegrate. Improvised rafts are really only suitable for still water travel, so if you feel the current pick up and see breakwater ahead then strike for the bank and walk the turbulent section.

TRAVEL

In the densest sections of jungle you will be lucky to make more than 1km (1000 yards) progress in a day, and plotting a straight course for that distance will be almost impossible. Of course, there are trails looping through the foliage, but an SAS patrol needs

to be very cautious of these as: a) they are frequented by locals; b) they offer perfect positions for enemy ambushes; and c) wild animals use them at night. Essentially, tracks are too open for SAS movement, so instead follow other routes through the jungle, such as rivers, hill ridges and the edges of fields. By picking salient features but travelling just off them your navigation will also be easier to follow on the map. Whichever route you take, you may well have to hack through lots of foliage. When doing so, chop with your machete or knife at an angle to the foliage, not straight down or across. Wrapping your hands in cloth can also protect you from attracting hundreds of tiny cuts from the grasses and twigs.

Of course, while doing this you still have to act like a military patrol. Never forget this – the instructors will be looking for someone who will retain focus on the operational duties despite all the other distractions. You will have been taught a new range of hand signals and SOPs for the jungle section of training. Learn them and use them confidently.

These are just some of the survival challenges facing the recruit in jungle training. It is undoubtedly an exciting, if somewhat bewildering, section of CT, and an entire book could be based around this section of training (a good book to read before SAS survival training is former SAS soldier John 'Lofty' Wiseman's *The SAS Survival Handbook*.) However, we must now look at a different aspect of survival, one based around a very distinct type of threat. This is the part of CT known as Escape and Evasion.

ESCAPE AND EVASION

Often at the end of the recruit's survival training, they face one of the most controversial and secretive sections of the training – Escape and Evasion (E&E). Current details of the course are not known, so all that is available is the recent testimony of those who have completed it. The picture that

RIGHT: Platform shelters are the defining type of shelter for use in jungle environments. A frame is made by lashing branches together and a platform built about 45cm (18in) above the ground. The platform keeps you away from ground-dwelling insects.

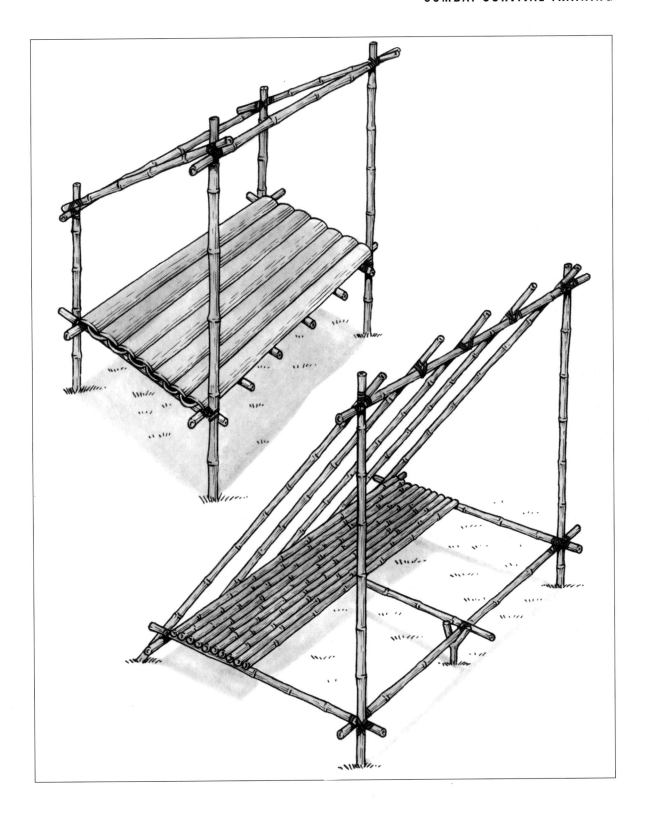

emerges is of a highly realistic exercise in pursuit and interrogation, during which the recruit must show the ability to evade capture and also to resist some incisive interrogation techniques. If he fails at any stage, then he is RTU'd.

So what do we know about the course content? In its outline it runs in roughly three stages. Firstly, the recruit is given training in evasion techniques, usually as part of his general survival training. Secondly, he and several other recruits have to conduct an evasion exercise, while being pursued by soldiers from other British Army units plus the instructors. Whether they are caught or not (the recruits have to evade until they reach a particular rendezvous) at the end of the exercise they will have to go into a Resistance to Interrogation exercise. This third section of the course is run by the Joint Services Interrogation Unit (JSIU) – responsible for interrogation training amongst all sections of the armed forces. Indeed, during the E&E the SAS recruit may find himself in a mixed team of soldiers which can include RAF and Navy pilots, another group highly desirable for an enemy unit to capture. During the Resistance to Interrogation exercise, the recruit will be exposed psychologically and physically to the type of treatment they could expect at the hands of captors, short of damaging physical torture. If they can get through a 24-48 hour period of this treatment without talking inappropriately to their interrogators, then they will have passed the E&E section of the course. Completing this section can also be the last element of CT, so the inspiration and motivation to do well is powerful. Pass this, and you could be a full member of the SAS.

We shall now look in more detail at the E&E content as it is known in the public domain, and pick up on the tips for successfully passing this element of the course. The pursuit section takes place in various wilderness locations across the UK, sometimes Exmoor, sometimes Snowdonia or the Brecon Beacons, other times up on the North York Moors.

RIGHT: Jungle locations are usually laced with waterways, and there is also plenty of wood for constructing rafts. All rafts should be tested in safe stretches of water before using them more ambitiously. They should be light enough to carry for short distances around obstacles.

The soldiers pursuing you – like a genuine enemy in pursuit – will use every human, canine and technological resource to find you

Whatever the location, you will have to spend up to three days on the run, all the while being pursued by British Army units. 'Pursued' perhaps does not do justice to the enthusiasm with which you will be tracked, as one SAS soldier points out:

'The hunter forces were f***ing serious! They had dogs, vehicles and helicopters, and a bloody good idea about where we were going to be. I think it was a company, or two companies, of infantry, and there were rumours that they got extra leave if they caught us so they were really up for it.'

This soldier's experience is typical. The soldiers pursuing you will not just search around vaguely in the direction you headed off in, but will instead – like a genuine enemy in pursuit – use every human, canine and technological resource to find you.

This is where you need the fundamental evasion skills. We can only touch on the basics of these skills here (you will be taught the full range of techniques prior to the exercise), yet these are amongst the most useful.

LEAVING 'SIGN'

'Sign' is the term used within British military forces for the evidence left on an environment by a person or unit passing through. There are various levels of sign: temporary and permanent, these being sub-divided into top sign (evidence above the ground) and ground sign (evidence at floor level). An example of temporary sign might be a footprint, as this will eventually disappear, and a dropped bottle would be an example of permanent sign. The key point for E&E is that you leave no sign for the pursuers to follow. Techniques for doing this include:

● Try to cross as many hard surfaces or water ways as possible where footprints will not be in evidence.
● Do not break foliage – the broken ends will

stand out clearly (particularly at night) and mark your route of passage.
● Never drop any litter – the clearest sign of human passage.
● Perform bodily functions in a place which will take away the evidence, such as in a stream.

By following simple precautions such as these, and generally being aware of any marks you might leave on the landscape, you will make the job of the pursuers that much harder. However, there is another type of sign about which you can do very little – that of smell. This is the type of sign picked up by tracker dogs, and you need special skills to evade those (see feature box).

Of course, evading dogs and not leaving sign are unimportant if you are simply spotted visually. Hence your movements should follow certain rules which you will have already learnt during your patrol SOPs. These include not using roads and trails, avoiding skylines, moving more at night, maintaining camouflage and not creating obvious light sources. Yet whatever your skills, the odds are still stacked rather heavily in favour of the pursuers (who will usually be aware of your techniques). Many soldiers are captured during the exercise – if you have had a decent period of evasion this does not constitute a failure. Even those who do make it to the end are instantly turned over for Resistance to Interrogation training.

RESISTANCE TO INTERROGATION TRAINING

Once captured by the staff of the JSIU, you are in for a rough ride. Though all present know that no physical harm will result, the interrogators will subject you to a barrage and range of physical punishments and inducements in their attempt to get you to talk. The scenario will also be realistic. Suddenly you will find yourself surrounded by men wearing foreign uniforms, carrying AK47s, and talking in foreign languages (when they are not screaming at you). Most former soldiers who have talked about

LEFT: A trooper sits by his fire in an arctic shelter constructed from poles and a waterproof sheet. Though it is very open, the key point is that it protects the soldier from wind chill. Heat from the fire will also be reflected from the rear of the shelter onto the soldier's back.

this experience report several common elements to the 'torture'. These include:

● Being bombarded by chaotic sounds ('white noise') at high volume from a stereo system for several hours.
● Left blindfolded outside in the cold and rain for hours, under total silence.
● Having a bag over your head for almost the whole experience, except when facing direct interrogation.
● Being placed in the stress position (stood spread-eagled against a wall with the weight heavily on your arms) for long periods, and being kicked back into position when you relax it.
● Being screamed at frequently.
● Being stripped naked, often being mocked by male and female interrogators.

These are the typical experiences of those on Interrogation training, yet it has been known to get tougher. Some soldiers have experienced having their heads plunged under water and only brought up when they are about to pass out. Some are stood

DEALING WITH DOGS

Military tracker dogs are trained to pick up your scent and follow its trail to the source. There is nothing you can really do to stop the dog detecting the scent, but there are several measures which can help you confuse the dog and its handler:

● Cross streams and rivers, or walk directly down them for a distance.
● Keep moving quickly – remember, the dog is only as fast as its handler.
● Occasionally make complex scent patterns which will confuse the dog. Do this by forming complex looping patterns over and around lots of different obstacles. By making some of these loops closed circles the dog will appear to its handler to be getting lost (as it well might) and start covering old ground.
● Stay down wind if possible.
● Split up a team of soldiers so that the dog has too many scent options.

with a bag over their heads in a field while cars perform skid manoeuvres around them. A famous 'training' incident involved a blindfolded recruit being tied to a railway track, and then the instructors feigning panic that the exercise was going wrong and a real train was coming (the train was actually passing on the adjacent track). We do not know what methods are used each new year, but these examples show how the recruit needs to prepare himself for a strange and alarming experience.

All the methods of 'torture' are designed to see how emotionally stable the recruit is under stress and also to soften him up for the interrogations. The JSIU interrogate the recruits to find out if they will state more than the customary name, rank and number. Interrogation technique is varied to disorientate the recruit. Sometimes you will face one apparently reasonable individual who will just try

All the methods of 'torture' are designed to see how emotionally stable the recruit is under stress and also to soften him up for the interrogations

to get you to talk about seemingly innocent things such as family, and life before the army (remember, for a trained interrogator every scrap of information is useful). Other times the interrogator will constantly fire the same questions again and again at you for hour after hour. The key behind this particular tactic is that you can find yourself slipping into a repetitive response, and then actually slipping into a revealing answer to any question which has not been part of the routine. Another time, you will get the classic good cop/bad cop routine – one talking fairly to you while the other screams invective straight into your face. Then you might be facing an attractive female interrogator, who will try to make you soften to her questioning, or claim that all your mates are talking and you're the only one left, so why keep going?

Whatever the combinations or the strategies employed by the instructors, they are looking for any methods which will make you open up to their questioning and become pliable. Either that or they

are looking for signs that you are genuinely breaking down under the pressure of the persistent and harrowing questioning. For your part, there are no hard and fast rules about how you cope with it – each person has their own technique.

However, there are several areas of mind control that can help you successfully resist these types of interrogation:

- Keep a sense of humour – give the interrogators ludicrous names in your head and try to see the absurd side of the situations you find yourself in.
- Be very observant with your senses. Imagine that you are trying to escape and look for any opportunities. This will give you a greater sense of control over the situation.
- Do not trust anyone, and never respond to even the most innocent seeming questions.
- Do not focus on any pain. Instead, try to keep your mind mentally active with a subject of your own choice.
- When you are not being questioned, try to rest and save your strength.

Research has shown that your response to JSIU training will differ in some ways from your behaviour if you were captured by a real enemy. Former SAS prisoners often recount how resisting dogmatically to questioning would just result in broken bones and other injuries. You need to avoid serious injury in order to survive and escape, so prisoners often prefer a subtle strategy of deception, if possible, giving the interrogators titbits of useless or out-of-date information until they lose interest.

However, the JSIU training does serve the SAS soldier well as an introduction to the strains of captivity. If the instructors observe that you have performed well, without a weakening of will or without giving away information, then you will have passed the course. If you do that, and have successfully completed everything else in CT, then you are now in the SAS.

CHAPTER 7

Life in the Regiment

There is no denying that life in the Special Air Service is fulfilling and frequently exciting, with the promise of real missions always on the horizon. However, training remains a key feature of life in the SAS, and it is necessary to update and constantly reassess the soldiers' skills.

'When you tell people you've been in the Regiment, they're asking you all about the warry stuff, blood and guts, and you give 'em both barrels, get their toes curling. But that wasn't what it was like for me. Since I've been out, the thing I miss is the sense of adventure, not the operations or the... glamorous bits. It's going into camp on Monday and not knowing where you're going to be three days later: Africa, South America, wherever. And you know you've got a set of mates who're going to be your mates for life: they'll die for you if they have to, and they know you'll do the same for them. Maybe that's the thing about the Regiment: it's f***ing hard to get on, and it can be f***ing rough when you're in it, but you're in it all together... whatever happens.'

There can be few better introductions to what life is like in the SAS than this, given by a former NCO of 22 SAS. The impression is of a unit in which

LEFT: Serving in the SAS means getting your hands dirty. Here an SAS team during the Gulf War confronts the problem of freeing their long-wheelbase Land Rover from a flood. Environmental conditions in the Gulf were one of the biggest dangers for the SAS, and cost some lives.

SAS PISTOL SHOOTING

The SAS place as much emphasis on pistol shooting as rifle and sub-machine gun technique, especially amongst the Counter-Revolutionary Warfare team, who might be engaged in covert actions where only pistols can be carried. Until recently, the 'double tap' was classed as the most efficient way to put an assailant or terrorist down (this technique was also taught to most police force units around the world). Simply put, the double tap involves putting two rounds into the torso of the opponent in quick succession.

The centre mass is chosen because it is the easiest to hit and contains almost all the body's vital organs except the brain. Though two rounds were at the time thought sufficient, today the target is the same but the policy is to continue firing into the target until he or she goes down. Studies had found that even after being hit by two rounds many determined people had still managed to return fire before succumbing, whereas a continuous stream of fire stops any response because the physical impact of the bullets prevents controlled movement.

adventure and comradeship are inextricably linked. According to its members, being a part of the SAS is amongst the world's most fulfilling vocations, a job which involves extremes of experience which few in the civilian sphere can contemplate. For those 10-15 people who survive Selection, this is the world they step into. The Regiment becomes a *de facto* extension of their family and they become part of an internationally respected unit.

This may describe the ethos of the SAS, but what of its day-to-day workings? When a new SAS soldier receives his cap and badge during an induction ceremony at Hereford – his Selection experiences and aches still fresh – what can he look forward to over the next year of his life in the SAS?

SAS STRUCTURE

To understand the life of a new SAS soldier you have to have a basic grasp of how the SAS is structured. What he will do and what he will become is very much governed by this structure and the place he occupies within it.

At the top of the SAS organization is the overarching umbrella – the Special Forces Group. Within this group fall the various Special Forces units of the UK: the Royal Marine SBS, the four SAS regiments (21 SAS TA, 22 SAS, 23 SAS TA, 63 SAS Signals Squadron TA – only 22 SAS is a regular force of the four SAS units, and thus it is our main concern here). 22 SAS then subdivides into its constituent parts. These are: Operations Research Wing; Headquarters Planning and Intelligence; Counter Revolutionary Warfare Wing; Training Wing; 264 SAS Signals Squadron; Attached Specialist Units; and finally, the units which are of most relevance to our new member, the Sabre Squadrons.

The Sabre Squadrons are the operational backbone of 22 SAS. There are five squadrons: A, B, D, G and R, the latter being a Territorial reserve unit. Each Sabre Squadron is then broken down further into a headquarters and four troops, each troop specializing in a different form of warfare. There are Mountain Troop, Boat Troop, Mobility Troop, and Air Troop. Taking the structure even further, each troop consists of a number of four-man patrols, each man within those patrols having a different speciality.

The new SAS trooper will generally first go into a Sabre Squadron, then receive training as part of a particular troop (though he will also receive training in all the other elements of SAS warfare). Furthermore, he will find or be given a Regimental speciality – anything ranging from linguist and medic to parachute specialist or artillery controller.

The irony of this for the soldier is that although he has just completed one of the longest and toughest training courses in the world, his training will recommence almost straight away in new areas. Indeed, training in the SAS never stops. The skills and techniques of SAS troopers need constant updating for new technologies, political situations, and conflicts. The big difference, of course, is that even though training continues the soldier is now ready to be used on actual military operations.

Life in the SAS is exceptionally fulfilling and exciting, with a lively social life and ever-present demands for foreign travel. Yet many soldiers also report how laid back life can seem in the SAS, especially to outsiders. Soldiers seem to be going about their business without the constant bark of orders

ABOVE: 'Pink Panther' was the first Land Rover adopted by the SAS. Its colour was inspired by the look of shot-down aircraft in North Africa in WWII, after sand and wind had stripped off paint and left a pink hue.

following their heels, and dress can seem decidedly casual. Yet the surface appearance of loose discipline is utterly false. What distinguishes the SAS from regular units is the knowledge that the only people who are in the Regiment are those who have the ability to exercise huge amounts of self-discipline without the need for external pressure. The operational nature of the SAS means that its soldiers have to be able to conduct missions far from the hierarchical structures of command, so if they cannot discipline themselves then their missions will be jeopardized from the start.

However, this does not mean that the new soldier can relax into whatever routine he likes. Training and operational demands are constant in the SAS, and the soldier needs to remain highly motivated to fulfil his daily duties. Furthermore, for the first year the soldier remains on probation – so even after the

rigours of Selection and Continuation Training he could still be out if he puts a foot wrong.

After induction, the first stage of an SAS soldier's career will be his placement in a particular Squadron. Once in that squadron he is then selected for training in a particular specialism. Sometimes the specialism will select itself on account of the soldier's pre-existing military or civilian skills. Thus, if he arrives into the SAS already able to speak two or three languages, then it is likely that he will become a linguist as one of his specialities. If he has a detailed knowledge of computers, then communications might be his avenue. The SAS will begin instructing the soldier in a skill which will contribute to the potency of a four-man patrol, the troop, and the Squadron to which he belongs.

SAS SPECIALISMS

As a new soldier to the SAS, the training you go on to receive in the Regiment takes you to the heights of expertise in all areas of military practice. Building on what you learnt in Selection and Continuation

ABOVE: Iraqi armour suffered horrific losses under the Allied air assault during the Gulf War. The SAS had teams of observers present in occupied Kuwait and Iraq dedicated to spotting enemy armoured movements and reporting their location back to Allied strike aircraft.

Training, the SAS makes you a fully operational part of the Regimental structure. Yet as well as having a broad education in the skills of an elite unit, the SAS will also work to make you a specialist within one or two particular fields. SAS soldiers are all masters of combat arts, yet it would be impossible to master every military skill required of the Regiment. Thus by training each soldier in one or more essential specialist fields, the SAS has all the talents it needs to put together any type of operation required of it.

The specialisms within the Regiment are numerous. Some are utterly conventional or expected – such as a demolitions expert or a communications specialist. But others are more tangential to regular army life – lock-picking specialists or linguists being examples. Whatever the specialism you fall into, the first stage in your new training life is dictated by the type of troop you are placed in: Mountain, Boat, Mobility and Air (we will consider the separate category of the CRW unit later). We will first look at the particular skills which come as part and parcel of each troop, then proceed to look at others outside these brackets.

MOBILITY TROOP

Mobility Troop is concerned with the effective movement of SAS units across land, primarily using wheeled vehicles. The emotional links to vehicle-borne combat units within the SAS is strong, as the unit was created around vehicular tactics in the deserts of North Africa in the 1940s. Indeed, the Gulf War saw SAS desert fighting units again taking to heavily armed vehicles once more, though the success of these applications was less influential than in World War II.

However, the Regiment still relies upon vehicular insertion and belonging to Mobility means becoming an expert in all aspects of motorized combat. 'All aspects' covers a surprising remit. Obviously, the Mobility Troop soldiers need to have a superb grasp of vehicular warfare, including techniques of defensive and offensive driving, attack formations of multiple vehicles, and the deployment of weapons. Operating so far from friendly logistics, the Mobility Troop soldiers also need to have considerable expertise in mechanics for the maintenance of their own vehicles. All troop soldiers have to spend time with the Royal Engineers learning these skills, demonstrating an ability to diagnose faults and make effective repairs where possible.

The same level of mechanical skill is required with the vehicles' onboard mounted weaponry. Most commonly, the vehicles are armed with GPMG or Browning machine guns, but these can be supplemented by tools such as the Mk 19 automatic grenade launcher and the MILAN anti-tank weapon

BELOW: SAS soldiers are trained to use MILAN anti-tank weapons. These huge missile launchers are rarely physically carried on operations – in the Gulf War the SAS used them mounted on Land Rovers. The MILAN can deliver an anti-tank warhead up to 2000m (6000ft).

system. Applying these systems from a fast-moving vehicle mount requires different tactical considerations to get results. Getting the vehicles into position for an attack is another problem. Vehicles do not have the covert presence which an individual soldier can achieve. Consequently, the soldiers of Mobility Troop have to learn new tactics of utilizing cover and surprise to get the most from their attack vehicles. This means changing the approach to map reading and route planning, and also mastering difficult techniques of insertion such as night driving. The training programme also tests the soldiers' learning in different environments, from the deserts of the Middle East to the wooded mountain wildernesses of the US.

The soldiers of Mobility Troop have to become familiar with one type of vehicle above all – the Land Rover. The Land Rover replaced the US Willys Jeep as the SAS standard vehicle in the 1950s. Its qualities are legendary – robust, all-terrain and all-weather capabilities, excellent traction on all surfaces, powerful braking ability, a high-performance engine, good load-carrying capacity – and it has found use in the hands of explorers and soldiers alike the world over. Many different versions of the Land Rover have been used by the SAS, but they currently use the Land Rover Defender 90 (short wheelbase) and Land Rover Defender 110 (long wheelbase). Also used occasionally is the Land Rover 130 – a special stretched-wheelbase version ideal for tasks such as ambulance work. An SAS customization of the Land Rover is the famous 'Pink Panther' version. The nickname comes from the colour scheme of the vehicle, an anaemic-looking pink which is one of the optimum colours for desert camouflage. Apart from the colour, the SAS have added smoke canisters and a spare wheel mounted over the front bumper, but have taken away the doors.

Another version of the Land Rover destined to arrive with the SAS soon is the Land Rover Special Operations Vehicle (SOV). This is actually a cross between a Land Rover and another SAS vehicle, the Light Strike Vehicle (LSV). The LSV was essentially a dune-buggy type-vehicle, a very light, fast automobile consisting of little more than a bodywork cage, a powerful engine, durable suspension, and heavy armament. Weaponry options were diverse, and could include 0.5in GAU-19 multi-barrelled machine guns, 40mm M19 grenade launchers, MILAN anti-tank missiles, FIM-92 Stinger surface-to-air missiles, and 81mm mortar. Like the Willys Jeeps, the LSVs were intended for recce and fast-attack operations, and along with US Special Forces the SAS put the LSVs into action during the Gulf War. It was here that the LSV's limitations became clear. Its small size meant limited fuel reserves and hence a

RIGHT: The Klepper canoe has been used by SAS and SBS teams engaged in coastal infiltration, and its last known major operational use was during the Falklands War. The Klepper is a wood-framed canoe and is collapsible to provide easy storage once ashore.

constrained operational distance (unless air supply could be arranged), and its potential for destruction was undone by a whole range of mechanical problems which hampered its operations.

The SOV is intended to add the reliability of the Land Rover to the firepower of the LSV. Indications are that the weaponry consists of two 7.62mm machine guns, fittings for LAW, MILAN or TOW anti-tank missiles, 51mm or 81mm mortar, and two Mk 19 grenade launchers. With an operational range of 748km (465 miles), the SOV will more than plug the gap left by the LSV.

The Land Rover series of vehicles are just one range of motorized transport in which the Mobility Troop will have to become experts. Trucks, standard British Army APCs, and various other small to medium size vehicles have to be mastered. By doing so the Mobility Troop provide an essential arm of support to the general operational flexibility of the SAS as a whole.

It is a tough and chilly part of SAS training. Indeed, of all the areas of dangerous SAS training, more men have been lost at sea than in any other environment

BOAT TROOP

While the SBS rightly commands much of the attention in the realm of elite amphibious forces, the SAS Boat Troop has similar capabilities and operational methods. However, a crucial distinction is that the Boat Troop focuses on amphibious insertion as a precursor to land-based operations while the SBS, though they can also perform this role admirably, tend to concentrate on staying water-based for as much of the operation as possible, if not all of it, for missions such as underwater sabotage.

However, it is evident that the two units cover much of the same territory – literally in fact as the SBS units are often based in Hereford. But this does not tend to result in any rivalry. Indeed, the soldier in Boat Troop will find himself in joint training exercises/operations with the SBS, and relations between these two outstanding elites are good.

(Reconnaissance and attack missions during the Falklands War and a combined response to a bomb threat aboard the liner Elizabeth II in 1972 are powerful examples of SAS and SBS collaboration.) Indeed, the SBS are frequently used by the SAS to conduct amphibious insertion of SAS teams into action.

Similar to the SBS, the Boat Troop has to be fully versed in aquatic insertion and survival. The training is tough – many hours spent in freezing ocean waters practising amphibious SOPs and covert insertion tactics. The result is an individual who has several key qualities. Boat Troop soldiers have an intense relationship with water, and so will be excellent swimmers and divers. The diving specialism will include the underwater sabotage of shipping and installations, plus the piloting of various one- or two-man submersible deployment vessels. They will also have to develop a knowledge and familiarity with the workings of a modern Royal Navy nuclear submarine, as these are occasionally used to take teams close to enemy coastlines and then deploy them underwater.

Ship boarding and assault is also handled by the Boat Troop. A merchant seaman officer recently recounted to me an experience he personally had of this form of deployment. His shipping company had already been approached by the SAS to ask permission for one of their teams to attempt to board a cargo ship passing through the South China Seas. Permission was granted, and consequently the crew of the ship were put on a full alert in the section of sea where the boarding was meant to take place. The sea around the ship was bathed in arc lights, and the crew stood watch at every corner of the ship. However, after several uneventful hours, the crew received a call over the PA system to go to the bridge. On arrival they found an SAS team already there, with the captain of the ship tied gently to a chair! To this day the officer does not know how they boarded the ship.

Such is the level of knowledge which the Boat Troop soldier comes to possess about all types of modern seagoing craft, military and civilian. This applies not only to boarding and taking ships, but also to piloting vessels of various types. The SAS has a whole range of vessels for its amphibious

operations. At the smallest end of the scale are man-powered craft such as the Klepper canoe. This is a two-man collapsible canoe 5.2m (17ft) long, and was first used in the 1950s. It provides an excellent method of totally silent insertion against enemy coastal and river positions, though it needs fairly stable sea conditions to operate effectively.

The subskimmer can switch between its role as a mini submarine and that of a surface-assault craft.

ABOVE: Soldiers of Boat Troop hit the shore during a training mission. They are using the Rigid Raider assault craft, a lightweight, glass-fibre reinforced plastic boat with a top speed of 37 knots. The front is designed to allow the boat to drive up on to shallow beaches.

The soldier changes between the two different functions by simply inflating or deflating of air compartments. Underwater, the craft relies on two 24v electric engines rather than the 90hp fuel

engine used on the surface. The soldiers using the subskimmer are not enclosed; they sit on the sub-skimmer wearing diving gear. Often the subskim-mer is used to put down on the seabed a short distance from the target or area of insertion, from where the soldiers complete the last few hundred feet of their journey by swimming.

Over and above these two craft come a sequence of much more conventional boats. One of the most frequently used is the Rigid Raider, a fast, hard-hull boat with a 140hp engine which can take an entire four-man unit to its destination with speed. Rigid Raiders were much used by the SAS and SBS during the Falklands War for deploying units into the Falklands' craggy, rocky coastline and conducting reconnaissance missions.

Amphibious operations are extremely tough by nature, so a key part of the Boat Troop training is survival at sea. This involves everything from marine survival navigation to how to fend off sharks, and it is a tough and chilly part of SAS train-ing. Indeed, of all the areas of dangerous SAS train-ing, more men have been lost at sea than in any other environment – evidently Boat Troop requires its men to be especially confident in the water.

AIR TROOP

Although all SAS soldiers are parachute trained in competent static-line jumping, the Air Troop adds new methods of parachute deployment, as well as other more unusual versions of air-deployment, to their repertoire.

The two parachute techniques which define the Air Troop soldier are those of HALO and HAHO. HALO stands for High Altitude Low Opening, indi-cating the content of this extremely risky form of parachuting. A HALO jump begins, as its name describes, at high altitude, often in the region of 10,000m (30,000ft). The high ceiling allows the deployment aircraft to stay out of the range of most surface-to-missiles, and at such heights it also pre-sents a tiny visual signature. HALO is designed to

LEFT: Paras of the elite Pathfinder unit are seen here performing a High Altitude Low Opening parachute jump. They keep an especially tight fall pattern – if they fail to do this, by the time they reach opening height they will be scattered well away from each other.

TREE JUMPING

Whatever parachute techniques are taught to the SAS recruits today, one technique omitted is that of tree jumping. Tree jumping was a technique of insertion developed during the Malayan crisis in the 1950s. The problem which faced the SAS in Malaya was that communist insurgent bases were situated deep within the jungle, many hours away by conventional patrol access.

Thus it was that two SAS troopers, Captain Johnny Cooper and Alastair MacGregor, proposed that the dense jungle foliage could actually permit the use of parachutes rather than deny it. Their idea was that the patrol members could be parachuted into the jungle and the thickness of the jungle canopy would actually arrest the descent before the parachute snagged on the branches and left the trooper suspended above the jungle floor. All that would remain would be for the soldier to cut or release himself, lower himself to

the floor using ropes, and then gather his equipment.

As treacherous as their proposal sounded, it did have the merits of offering surprise and the most direct form of deployment against the communist bases. Indeed, the first operation saw the drop of 54 troopers into the jungle in February 1959. Amazingly, there were few injuries and thus 'tree-jumping' appeared to have an operational future. However, this first jump was soon confirmed as more of an aberration than a rule. On subsequent jumps the losses through impact injuries amongst the tree became unsustainable. During one tree jump in January 1954 three SAS men actually died performing a jump. It was obvious that tree-jumping was untenable, and it was abandoned. The problem of access to the communist bases was resolved later in the conflict when helicopters became a viable method of insertion.

get the SAS parachutist from this high altitude to the ground in as fast a time as possible, also reducing the amount of time he spends conspicuously swinging in the air on the end of his parachute.

To do this involves a long period of fast free fall, with actual parachute opening left until the soldier is less than 1000m (3280ft) above the ground. The hazards of HALO jump are heightened by the fact that many HALO operational jumps will take place at night. Falling for about six miles through the night air and clouds is a disorientating experience. After a minute or so of freefall many HALO parachutists report a sensation of being stationary, simply suspended on a buffeting column of air. This presents obvious dangers in terms of knowing when to open the parachute – the ground is, after all, rushing up at the parachutist at around 250km/h (160mph)! Most HALO parachutists use automatic parachute release mechanisms and all have altimeters attached to their wrists, but the danger is still very real. Another problem stems from the environmental difficulties. At around 10,000m (30,000ft) the temperatures are far below zero, especially when combined with the prodigious wind chill of those altitudes and the breathing difficulties which result from falling through the thin air.

The parachutist is heavily kitted out in a specially-insulated jump suit with breathing apparatus, but ice can still build up rapidly on visors and altimeter screens. As the free fall descent relies on the parachutist holding a good body position – especially if he wants to land in the same place as his team – then struggling to clear ice can cause a real problem. The jump will also come after a chilling period spent on the drop aircraft (usually a Hercules) at high altitude. Soldiers cannot jump from a pressurized aircraft, so the aircraft is depressurized and the soldiers waiting to jump must breathe from an onboard oxygen supply before switching to their own bottles for the jump.

HALO has an equally risky partner, HAHO, which stands for High Altitude High Opening. The jump altitude for HAHO parachuting is about the same as HALO, yet the parachutes will be opened at about 9,000m (27,000 ft). Both HALO and HAHO techniques use the specialist type of parachute known as a flat canopy ramair parachute. This is a rectangular parachute with aerodynamic properties akin to an aircraft wing. These properties make the parachute acutely steerable, something which gives both types of jump greater precision in the landing. But for the HAHO jump this controllability has an added significance, in that it allows the parachutist to 'fly' up to 80km (50 miles) from the jump point to his drop zone. HAHO is employed when it is preferable for the drop aircraft not to enter a certain air space (usually for enemy radar or air defence reasons), but it is a difficult technique. No matter how competent the team, it will be extremely difficult for them to stay together during HAHO descent over long periods, so it is often better for the insertion of individuals. Furthermore, the parachutist spends a long period in the air, exposed to the harsh air temperatures and wind, conditions in which his equipment may suffer damage.

HAHO and HALO jumping are the defining insertion techniques of the Air Troop, but they will also receive some training in other forms of air mobility. For example it is useful to have one or two trained

The climbing training is exhaustive, and covers a range of surfaces from bare rock face to glaciers and even frozen waterfalls

aircraft or helicopter pilots amongst an SAS team. Air Troop soldiers can also find themselves training in more unusual craft, such as hang-gliders, microlights, and even Powerkites. Powerkites are basically large powerful kites used to drag people along the ground at speed on skis or in unpowered wheeled vehicles. Power kiting has some interesting implications – ski troops, for example, could be towed up a ski-slope by the powerkite – but it remains to be seen whether there is a more useful operational application for it. Whatever the tools, the Air Troop remains an invaluable tool for the SAS, enabling it to place teams of soldiers with great precision far behind enemy lines and with the minimum possibilities for discovery.

MOUNTAIN TROOP

Mountain Troop contains the mountain warfare specialists of the SAS. Mountainous training has already been encountered to a certain degree during the process of Selection. However, the Brecon Beacons, hostile as they can be, do not compare to many of world's more elevated mountain ranges.

ABOVE: **SAS parachute wings. This coveted badge is obtained during Continuation Training by those who have completed the requisite eight jumps, including one over water and one at night. More advanced techniques are taught after CT to the members of Air Troop.**

Mountains present distinct challenges for an SAS team. They offer extremes of temperature, wind strength, low-visibility, geological dangers, and hostile terrain found in few other places on earth. Hence the Mountain Troop spends much of its time training in the arts of mountain survival.

All SAS units seek out other elite units around the world which have something to teach them about the skills they are trying to master. Mountain Troop is no exception to this rule. Four units in particular are utilized to help establish Mountain Troop's skills: the German Army, the Royal Marines, the Italian Alpini, and the US Marine Corps.

A particularly close relationship is struck between the Mountain Troop and the Royal Marines. The Royal Marines have extensive commitments to Norwegian deployments, and consequently they have developed world-class expertise in mountain and arctic warfare. There are two courses run by the Royal Marines which attract would-be members of the Mountain Troop. The first is the Arctic Warfare Training Course conducted in Norway. This is a three-week course of intensive instruction in arctic survival. The SAS soldiers who find themselves on this course have much to endure. Tests include living out of four- or 10-man snow holes for several days in the freezing Norwegian climate, ascending glaciers, training in avalanche rescue, and cross-country skiing. The latter involves the dreaded 'ice test', in which the soldier will have to ski through a hole in the ice into freezing waters, then take off all his equipment and claw his way out using only his ski sticks. In addition to survival, the course also instructs soldiers in

The soldier receives skills which not only serve him well as an operational soldier, but which he can also take into recreational life beyond the Regiment

the techniques of winter warfare, expanding the SAS troopers' knowledge of matters such as winter camouflage, patrol techniques in snowy landscapes, and how to conduct successful winter ambushes. The course is a thorough introduction into the arts of arctic warfare, yet the Royal Marines offer a more advanced programme run by their Mountain & Arctic Warfare Cadre. This is actually an eight-month course – a considerable slab of a SAS soldier's time with his new unit. It is also incredibly tough even by SAS standards. Most of the people who attend the course are already hardened RM or SAS soldiers, yet despite this fact the failure rate runs somewhere in the region of 60 per cent! The

training is a far more intensive version of the Arctic Warfare Training Course, but offers a more comprehensive treatment of mountain warfare in general, looking at mountains beyond those simply covered in snow and ice. There is intensive climbing instruction and rigorous teaching of combat skills in mountain settings. Skiing is particularly emphasized, and the soldiers must pass the Military Ski Instructors Course.

An alternative, or an addition, to Royal Marines training, is given by the German Army. The German Army Mountain and Winter warfare School is set in the heart of the Bavarian mountains at Luttensee, near Mittenwald. The five-week course offered by the school is split between two locations, the Bavarian Alps and the French Alps. In both locations the soldiers are put through intensive survival and climbing instruction. The survival skills are quickly learnt – they have real value straight away in the freezing conditions as the soldiers are left to camp out on the exposed mountain sides. The

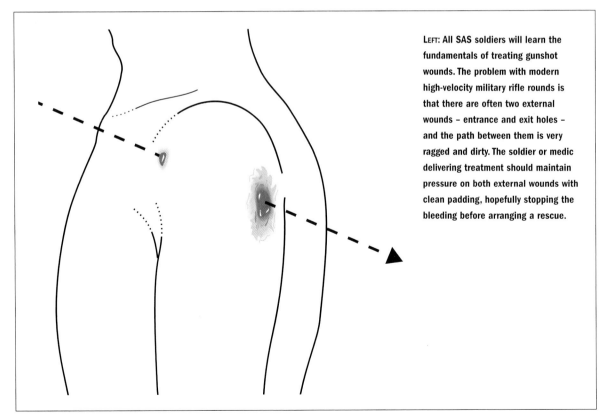

LEFT: All SAS soldiers will learn the fundamentals of treating gunshot wounds. The problem with modern high-velocity military rifle rounds is that there are often two external wounds – entrance and exit holes – and the path between them is very ragged and dirty. The soldier or medic delivering treatment should maintain pressure on both external wounds with clean padding, hopefully stopping the bleeding before arranging a rescue.

soldiers also quickly learn to assess the symptoms of altitude sickness, and attempt to adjust to this potentially fatal condition.

The climbing training is exhaustive, and covers a range of surfaces from bare rock face to glaciers and even frozen waterfalls. This training aims to give the soldier great confidence in tackling any climbing operation. The intensive course designs the training around the unique demands of soldiers in combat, with a faithfulness to the type of additional gear carried which would not be present on a civilian climber. Following climbing training, the soldiers will then proceed to a six-week ski training course at the end of which the soldiers are obliged to pass a German Ski Association's Instructor Test regardless of their starting level of expertise.

The Royal Marines courses and that offered by the German Army are just two of the potential courses available to the new Mountain Troop soldier. Other courses are also provided by the Italian Alpini mountain troops and the US Marine Corps' Mountain Warfare Training School. Both types of course offer an excellent grounding in mountain warfare and survival, although the Alpini course possibly has the advantage in that many of its personnel are born, bred and work in the mountains, and thus understand the environment with a subtlety often absent in other units. The US Marine Corps' course – held in the Nevada mountains – is also very useful, especially for its high-altitude training as well as its cross-country skiing instruction.

The Mountain Troop will provide Special Air Service response to any mountain combat environment, whether it is in an arctic environment or not. The extensive training to become a member of the troop is long and involved, but the soldier receives skills which not only serve him well as an operational soldier, but also which he can take into recreational life beyond the Regiment.

COMBATING TERRORISM: CRW

Of all the specialisms within the Special Air Service, it is those expressed by the Counter Revolutionary Warfare (CRW) wing of the Regiment which spring most readily into the public mind. This is primarily because of the high visibility of the Princes Gate action, which brought the SAS to public attention

and also gave them a vivid visual representation of the Regiment in those black-clad figures abseiling down the side of a building.

Almost all SAS soldiers spend some time with the CRW, and it is one of the most exciting and mentally stimulating aspects of SAS life. Each and every day the Killing House resonates with the bangs and crashes of SAS troopers practising their hostage-rescue actions. But beyond this limited view of the CRW unit, what is the reality of their training and composition, and what could you expect to learn within their ranks as a new soldier?

First, a brief introduction to the history of the CRW Wing. Though the massacre of Olympic athletes

STUN GRENADES

Stun grenades are one of the most important weapons in the SAS arsenal when it comes to hostage rescue. Because hostage-rescue missions are faced with the physical proximity of hostage-taker and hostage alike, the use of conventional fragmentation grenades would be unacceptably indiscriminate. The stun grenade, therefore, is not designed to kill but rather incapacitate, buying the SAS soldier a vital few seconds while the victim recovers.

The standard G60 stun grenade (manufactured by Royal Ordnance) emits a single defining bang of 160 decibels and a blinding light of 300,000cd. Anybody in the same room as this concussion will usually be incapacitated for up to ten seconds (usually around five), and this will give the SAS trooper time to enter the room and assess the situation without instantly meeting return fire.

The G60, however, is one of only a range of stun grenades. The German firm PPT makes a stun grenade, the Type A, which features a similar loud bang as the G60 but has a light which radiates for a dazzling 15 seconds. Their Type B reverses this priority and gives out eight loud bangs, while the Type C is all light, but no bang. Another company, Brocks Pyrotechnics, makes a similar range of grenades with various combinations of light, noise and also smoke emission.

Stun grenades are a very useful tool, and their regular use in SAS training means that the users themselves become accustomed to the powerful sensations of these vital weapons.

at Munich in 1972 certainly galvanized the creation of the CRW Wing, the SAS was fully involved in counter-terrorist duties some years before this date. The explosion of sectarian and anti-military violence in Northern Ireland in the late 1960s and the need to protect heads of state on foreign visits had led the SAS to be fully engaged in training for domestic and international terrorist incidents. Yet while this state of readiness existed without clear distinction from the Regiment's other duties, the Munich massacre led the SAS – like so many international military and police units – formally to establish a specified unit for dealing with the new terrorist threats. Created in 1973, this was called the Counter Revolutionary Warfare Wing.

Since 1973, the CRW Wing has become an integral part of SAS life and operations. Today, the CRW exists in a state of permanent training and constant readiness. Troops pass in and out of the CRW on a rota basis, but while in the unit they will belong to two teams, A team and B team. One team will generally be conducting training exercises within Hereford, while the other will be on external duties or training. Whatever the case, the CRW remains on 24-hour standby, ready to respond to any domestic or foreign emergency that warrants their attention (requests for help with domestic situations usually come from the Police, often with government recommendation or intervention).

Life in the CRW Wing is more diverse than is often imagined. Hostage-rescue is certainly a primary element of the CRW training, and this is conducted almost constantly in the Killing House. Though recruits going through Continuation Training will have been introduced to Killing House tactics, the CRW Wing will take their skills to a level of fluid expertise. Day in and day out, the soldier will find himself charging round the Killing House's smoke-filled rooms, responding each time to different set-ups and scenarios, the instructors throwing every conceivable problem in his way to see how he and the team will respond.

Visits from VIPs are also common, particularly politically sensitive members of the government (including prime ministers) and members of the Royal Family. On these visits the VIPs actually take part in the type of hostage-rescue scenarios which

ABOVE: An SAS team prepare to storm the Killing House during hostage-rescue training. The soldier on the right is armed with an Arwen 37mm grenade launcher with a five-round revolving magazine. He will fire CS gas into a room before others in the unit enter.

could face them in reality. Many members of the Royal Family, for instance, including Prince Charles and Princess Diana, have sat in one of the rooms and kept incredible composure while SAS units

On these visits the VIPs actually take part in the type of hostage-rescue scenarios which could face them in reality

passing on their own skills to the Royal Protection Group, who now attend Hereford as a matter of course in their training.

Training in counter-terrorism is also conducted in many other locations than the Killing House. The Hereford base contains aircraft and train carriages for practising assaults in these environments. There is also a two-storey building known as 'the embassy' for use in practising advanced multi-level building assaults. However, the CRW soldiers will also find themselves frequently on the road away from Hereford. The community of worldwide hostage-rescue and counter-terrorism organisations has little of the international rivalry that exists between other types of military unit. Consequently inter-training between the world's best HRUs (Hostage Rescue Units) is commonplace. For the CRW soldier this may mean a trip to Germany, for example, to train with GSG 9 or the US to train with Delta Force. However, because of the revered status of the SAS, the CRW soldiers are more likely to be abroad instructing such units rather than receiving instruction. In actual operational incidents, SAS advisors have frequently flown rapidly to various international destinations to provide constructive advice on the situation in hand.

Hostage-rescue is the most visible edge of CRW, yet the new soldier might be surprised at the range of other operations for which he needs to be trained. Counter Revolutionary Warfare is an umbrella term which covers a multitude of demands, some of which are more in line with conventional soldiering. These include pursuit operations against insurgents in rural areas, ambushes, surveillance of terrorist organizations, conducting hearts and minds operations in areas affected by terrorism, training civilian guerrilla forces. Perhaps the least visible task of CRW is the infiltration of terrorist organizations themselves as agents. Not everyone is suited to this role, as it usually requires a sophisticated mastery of local language, customs

burst in and spray the 'terrorists' with bursts of SMG fire. The purpose is deadly serious, for if such VIPs were ever taken hostage they would have a good idea of how to behave to assist the SAS should a hostage-rescue mission be launched. Protecting VIPs is firmly amongst the CRW's requirements, and they will be trained in all manner of close-protection duties. This includes defensive driving techniques and handling crowd incidents, as well as

LEFT: Abseiling is one of the combat deployment methods which arrived with the creation of hostage-rescue forces in the 1970s. The SAS are experts, as they proved during the breaking of the Iranian Embassy siege in 1980, when squads abseiled from the roof.

and beliefs, and the ability to act out the part with utter conviction. Those chosen for this type of role are very few (it takes exceptional courage and mental endurance to be able to cope with the pressure) but those who have shown some ability with

accents and languages are often the most favoured candidates.

Naturally, the sheer diversity of roles within the CRW means that the new soldier has to become acquainted with a whole new range of equipment. Just the hostage-rescue remit alone requires adjustment to weaponry such as the H&K MP5 and the array of assault tools, from ladders to stun grenades. However, CRW training is doubtless one of the most satisfying periods of an SAS soldier's career.

ABOVE: These illustrations depict a selection of the correct firing positions – as taught to the SAS – for urban combat. Note how obscuration works on two axes, vertical and horizontal. This prevents the soldier's head from making a sharp silhouette against the straight line of, for example, a roof.

OTHER SPECIALISMS

The list of individual specialisms held by the SAS is not definitive, and one man's unique interest in, say, arable farming can become a valuable specialism in itself for hearts and minds operations in rural Africa. (During operations in Borneo, soldiers were trained in basic veterinary skills to help the local people maintain healthy livestock.) Yet there are certain skills which the SAS commonly fosters in its men. Here we will look at some of the most salient ones, noting what the specialism involves and how it contributes to the portfolio of the Regiment.

MEDIC

Having soldiers who are medically trained is vital to SAS operations. Far behind enemy lines, an injured soldier might have to wait days before rescue comes to him, and thus he will absolutely depend upon the quality of first aid he receives there and then. All SAS soldiers have basic combat first aid, but within every four-man patrol will be a soldier who has trained in field medicine to an advanced degree. What the SAS look for is someone with a high degree of composure as well as an ability to handle the casualty's mental and physical state.

The job of a combat medic is very different to that of a civilian medic. Whereas a civilian medic will generally treat a casualty in a protective or helpful environment, the SAS medic may have to patch someone up while the enemy is working hard to ensure that no one survives, let alone the casualty. Also, the combat medic will have to treat the wounds that come with military weaponry much more frequently. A hospital doctor in London might see the massive cavitation injury from a high-velocity rifle bullet once in his career, yet the combat medic could treat dozens in the space of an hour. Similarly, what civilian doctor experiences the effects of phosphorus grenades, the particles of phosphorus in the skin reigniting in contact with the air as surgery commences? To gain a more realistic training, SAS medics are now commonly sent

Left: A trooper taps in a message into his PRC 319 communications set. The position of signaller is not just about maintaining radio communications, but also involves developing and breaking codes, commanding air and artillery strikes, and coordinating units in assaults.

to major hospitals abroad that see a lot of trauma from guns. At a city hospital in Johannesburg, South Africa, for example, the doctors can expect to see two or three firearm-related injuries every single day, so it acts as a magnet for combat medics all over the world.

But medics are not just useful for the treatment of combat injuries. They also have a vital role in the SAS's bridge-building hearts and minds campaigns. Medics in Malaya, for example, would provide midwifery care to the local women or help with inoculating the indigenous population against various tropical diseases. In later operations, SAS medics have helped teach local people about the essential principles of sanitation and hygiene, thus reducing the number of deaths from infectious illnesses such as cholera and TB.

It is evident that the medic is an invaluable specialism within the SAS, and one with a surprisingly broad operational relevance.

Linguist

The same can be said concerning the SAS linguist. All soldiers who join the SAS have the opportunity to undergo language training at the British Army School of Languages at Beaconsfield. Most accept this opportunity (they are encouraged to do so) as the unit profits tremendously from individuals who can communicate with local people abroad.

The reasons for this are obvious. Hearts and minds operations, for instance, are usually impossible without a grasp of the local language, not only for the basics of human communication but also because soldiers who have other languages tend to grasp cultural subtleties better than those who don't have similar linguistic skills. Also, interrogation becomes impossible if there is no common communication, and the combat medic's job is made easier if there is someone there to explain treatments given to local people.

The Army language course does not necessarily teach a soldier to be completely fluent. However, it does furnish him with the essential grammar and vocabulary to conduct conversations and read the local language (often a vocabulary of around 500 words is all that is needed for a decent conversational ability). In terms of the world's most useful

RIGHT: PE4 is a standard plastic explosive used by the SAS, others including RDX, PETN and SEMTEX. Seen here is a stick of PE4, a coil of detonating cord, and an electric detonator buried in the explosive. Only a detonator will set it off – set fire to it and it will just melt.

languages, the three most prevalent are English, French and Spanish (Spanish covers the remit of Latin America while the French language is still spoken in many regions of Africa and Southeast Asia). However, the Regiment also likes to foster skills in the languages of areas of particular operational interest, such as the Middle East. For those soldiers who manage to attain fluency in a particular language, it is possible that undercover or long-term work abroad might beckon.

Signaller

If Signaller becomes your speciality within the SAS, this does not mean that you will spend your time chained to a radio set. Signalling covers a multitude of communications technologies and tactics. At its most predictable level it means ensuring that a

The Army language course does not teach a soldier to be fluent, but it does furnish him with the essential grammar and vocabulary to conduct conversations and read the local language

unit's communications with important third parties work effectively and consistently in all terrains and climates. This is a challenge in itself, and involves a reasonable grasp of science and the ability to handle technically advanced communications systems such as the SATCOM (Satellite Communications) device and global positioning systems.

Yet in addition to keeping important lines of communication open between friendly parties, the signaller within the SAS will also become an expert in all surveillance technologies – as well as the techniques of using and breaking secret codes. This may include the more so-called 'cloak and dagger' type of operation, such as learning the ability to plant minute listening devices in clothes, cars or

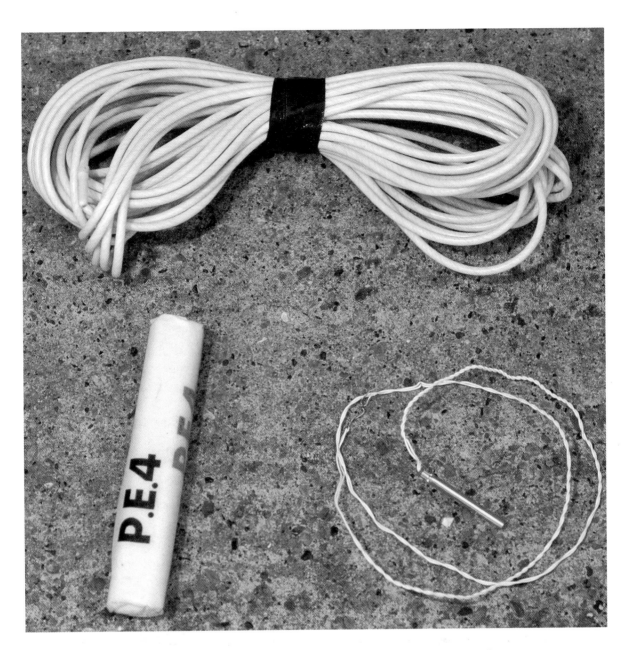

hotel rooms (such signalling skills have been used to great effect over the years to gather information about terrorist activities in Northern Ireland), to more overt actions such as jamming enemy broadcasts or establishing laser-designator communications with an incoming flight of attack aircraft. As is the case with all SAS specialities, the operational remit is especially broad.

DEMOLITIONS EXPERT

The demolitions expert has one of the more spectacular training periods of any SAS specialist. During the eight-week demolitions course he will be versed in the arts of blowing up and demolishing an amazing variety of structures and vehicles, and will actually get to practise his new-found skills within various British Army ranges.

However crude the title 'demolitions' may sound, it is in fact a speciality of some subtlety and sophistication. Not only must the soldier be trained in the physical properties of all the various explosives available, he will also undergo training in the science of constructions such as bridges and high-rise buildings, and of military vehicles such as main battle tanks and ships. Demolitions is the science of exactly matching means to structure (erring on the destructive side of the equation), and it takes some knowledge of how structures resist forces to know how best to apply demolitions materials. For example, a single-span suspension bridge requires the charges to be put in a different place than a cantilever bridge, and an armour-plated tank must be targeted at its vulnerable points otherwise explosives are likely to have little significant effect.

SABOTAGE SKILLS

Yet demolition is not just about explosives, though that is a large part. The demolitions specialist is also trained in the art of non-explosive sabotage, particularly industrial sabotage. This enables the soldier to train indigenous saboteurs, much in the same way as SAS units assisted the French resistance in World War II. Knowing how to make an internal combustion engine fatally, but inconspicuously, overheat, or how to destroy a computer motherboard are skills just as vital as blowing up an ammunition dump in today's complex wars. To this end, training involves the soldier making many guided tours around industrial plants such as oil refineries and chemical works, to learn from the experts what actions would most disrupt production.

Finally, the demolitions specialist is becoming more and more involved in remote destructive technologies, particularly in the use of laser designating systems for airstrikes or coordinating artillery bombardments.

To be an SAS soldier is to be a specialist as well as a superb generalist, which explains why so many SAS soldiers go on to do well in civilian vocations after their military careers. Yet during those careers, each SAS soldier will hope to achieve the ultimate expression of his learning, an actual operation.

No one can say how, when, or where an SAS soldier joining the Regiment today will serve. However, from the accounts of former soldiers, it seems that if you spend a three-year period in the SAS, operational deployment is almost certain. For a start, we may never know exactly which operations the SAS have been involved in. But from those oper-

RIGHT: A demolitions expert places a ball of plastic explosive against a railway line prior to blowing the track. Plastic explosive can burn at up to 8500mps (27,887fps) and only 15kg (32lb) of modern RDX has the same explosive force as 90kg (190lb) of C4.

ations we do know about, it is clear that there have been few periods since World War II when the Regiment has not been operationally active. Deployments in the past few years have included Sierra Leone, Yugoslavia and most recently Afghanistan, and world politics are already offering many possibilities for future operations.

Deploying small units of elite forces is much more cost-effective for a budget-conscious British government than deploying large-scale forces. This may go a long way to explain why investment in the Special Forces has remained consistently high despite cutbacks across the broader military.

The call to action can arrive at any moment for a soldier in the SAS, and once it does he will be facing the moment of truth in his military career. However, as we have seen during the course of this book, if any military operation does come up, and the soldier finds himself whisked halfway across the world ready to fight, he will go with the backing of the most thorough and superlative military training of any unit in the world.

SAS Chronology

November 1940
The recently formed No 2 Commando is renamed No 11 Special Air Service Battalion after parachute training.

July 1941
L Detachment, 1st Special Air Service Brigade is formed under the command and the inspiration of David Stirling, the founder of the SAS.

16–17 November 1941
L Detachment launches its first action against the Sirte airfield, North Africa. The operation is a disaster.

12 December 1941
L Detachment launches its next operation against Tamit airfield. This time the operation is a success, with 24 Axis aircraft destroyed.

14–26 December 1941
L Detachment launches numerous more raids against Axis airfields in North Africa, resulting in over 70 enemy aircraft destroyed.

23 January 1942
L Detachment and Special Boat Section attack Axis-held port on the North African coast. Though no shipping is found, warehouses and petrol storage tanks are destroyed.

8 March–26 July 1942
L Detachment's programme of attacks against enemy airfields in North Africa continues. Up to 200 aircraft are destroyed, and hundreds of enemy personnel killed.

October 1942
L Detachment is renamed 1st SAS Regiment.

January 1943
Lieutenant David Stirling is captured during operations in North Africa, eventually being confined in Colditz Castle prison. Shortly afterwards 1 SAS is renamed the Special Raiding Squadron and the Special Boat Section becomes a separate unit.

May 1943
2 SAS is formed under the command of David Stirling's brother, William Stirling. It conducts unsuccessful reconnaissance and prisoner snatch operations on the island of Pantelleria and Sardinia respectively.

10 July 1943
40 men of 2 SAS capture an enemy gun battery as part of the Allied landings on southeast Sicily. Two days later the SAS launch an unsuccessful action in northern Sicily as in attempt to disrupt German communications on the island.

September 1943
Special Raiding Squadron and 2 SAS launch attacks on German communications and railway lines in both northern and southern Italy, with a high degree of success.

2–6 October 1943
61 men of 2 SAS rescue 50 Allied POWs between Ancona and Pescara in southern Italy.

3 October– 10 August 1943
2 SAS conduct extensive operations during the Italian campaign, striking at enemy communications, railway lines, bridges and airfields.

January 1944
The SRS is reformed into the SAS Brigade, which now consisted of two British units (1 and 2 SAS), two French units (3 and 4 SAS), and one Belgian unit (5 SAS).

June 1944
From D-Day, 4 SAS are initially deployed in Britanny, France, organising local resistance and harrassing enemy movement. 1 SAS are divided between positions west of Dijon, Carentan on harrassment and deception operations. Throughout the French campaign the SAS are used in a wide variety of attack, reconnaissance and resistance roles, and are rarely out of action (key actions are listed below).

16 July–7 October 1944
3 SAS conducts Operation Dickens - 65 men launch actions against the German railway system in western France, killing 500 enemy soldiers and destroying 200 German vehicles.

25 July–15 August 1944
Seven men of 2 SAS, commanded by Captain

Lee, make an unsuccessful attempt to kill Field Marshal Rommel at Rambouillet.

13 August–26 September 1944
107 men of 1 SAS are deployed to assist Allied airborne landings in the Orleans Gap, inflicting heavy casualties on the German forces.

29 August–14 September 1944
4 SAS around Bourges, central France, attack columns of retreating Germans. Many Germans are killed by the SAS strikes, and the unit takes 2500 prisoners.

2 September 1944
5 SAS begin to conduct operations in Belgium, beginning with attacks on enemy communications at Liege-Aachen-Maastrict.

16 September 1944–14 March 1945
Five men of 5 SAS operate in a reconnaissance role around Arnhem, Holland, reporting effectively on German positions and V2 sites.

27 December 1944–15 February 1945
2 SAS conducts Operation Galia in northern Italy, between Genoa and La Spezia, attacking German troops and liaising with the Italian partisans.

27 December 1944–15 January 1945
5 SAS work in support of British forces resisting the German armoured offensive through the Ardennes.

25 March–3 May 1945
1 and 2 SAS begin operations using heavily armed jeeps within the German fatherland itself, fighting east of the River Rhine to Kiel.

April 1945
2 SAS operate with partisans against enemy garrisons and communications on the Italian Riviera, with success.

6 April–6 May 1945
1 SAS led by 'Paddy' Mayne takes heavy losses during reconnaissance missions for the Canadian 4th Armoured Divisions in northwest Germany, around Oldenburg.

12 May–31 August 1945
1 and 2 SAS and HQ SAS Brigade conduct mass operations (845 men) to disarm 300,000 German troops who have surrendered in Norway.

October 1945
The SAS Brigade is officially disbanded.

1947

The Territorial Army SAS Regiment is formed, commanded by Lieutenant-Colonel B.M.F. Franks.

1950–1951

The Malayan Emergency begins in 1948. Under the recommendation and leadership of Brigadier Mike Calvert, the Malayan Scouts - a counter-insurgency unit - was formed in 1950. They begin daring and effective raids into the Malayan jungle against the insurgents. By the second half of 1951, the unit was called the Malayan Scouts (SAS) - part of the make-up of the Scouts was from the 21 SAS territorials. By the end of 1951 the unit was entirely renamed 22 Regiment, Special Air Service.

February 1952– February 1958

The new 22 SAS conducts extensive operations against the communists in Malaya. By the late 1950s, the action of the SAS and the wider British forces have led to the defeat of guerrilla forces in Malaya.

1955

New Zealand SAS (NZSAS) is formed and joins the British SAS in Malaya.

1957

NZSAS is disbanded with the end of hostilities in Malaya.

November 1958– January 1959

SAS units (A and D squadrons) are deployed to northern Oman to fight against rebels. On 27 January 1959, the SAS take the rebels headquarters on the Jebel Akhdar plateau.

1959

The NZSAS is reformed, and the British 23 SAS (TA) is created.

1960

Hereford becomes the permanent base of the SAS.

1962

The NZSAS go to Thailand to cooperate in military assistance duties with the US forces. This involment would lead Thailand into the Vietnam War.

January 1963

A Squadron SAS arrives for operational duties in Borneo, beginning three years of SAS combat tours in the country fighting insurgents and Indonesian Army forces who are resisting the political formation of the Federation of Malaysia.

1963

The NZSAS are renamed 1 Ranger Squadron NZSAS.

4 September 1964

Australian Special Air Service (SASR) is formed.

April 1964– November 1967

The SAS conduct over three years of operations in the Radfan area of Aden against anti-British forces, though their commitment ends with general British withdrawal from Aden in November 1967.

July 1964

264 (SAS) Signals Squadron formed as an additional unit based at Hereford.

June 1966

SASR deployed to Vietnam. The NZSAS would join them in 1968.

1969
The SAS begins an involvement in the Northern Ireland conflict which would last until 1994.

July 1970
The SAS begin deployment in Oman. Over the next six years they help the government of Oman defeat communist revolutionaries.

1971
SASR and NZSAS withdraw from the conflict in Vietnam.

19 July 1972
A nine-man SAS team successfully defends itself against a large guerrilla force in the Battle of Mirbat, Oman.

1974
The SAS form the Counter Revolutionary Warfare (CRW) Wing as a unit dedicated to fighting international terrorism.

10 July 1978
The SAS mistakenly shoot a schoolboy, John Boyle, in Northern Ireland.

5 May 1980
The SAS conduct their most famous action, Operation Nimrod, freeing hostages from the Iranian Embassy on Princes Gate, London.

April–June 1982
The SAS is engaged in extensive operations in the Falklands War. This includes participating in the recapture of Grytviken and South Georgia, the destruction of enemy aircraft on Pebble Island (14-15 May), and the assault on Mount Kent (30 May). However, the SAS suffers one of its worse tragedies, when 18 members of D Squadron die in a helicopter crash (19 May).

8 May 1987
The IRA's East Tyrone Brigade is killed by an SAS ambush at Loughall, Northern Ireland.

6 March 1988
Three IRA terrorists are shot dead by an SAS team on the island of Gibraltar.

1989
SAS soldiers are sent to Columbia to train law-enforcement and military units in their war against drug production and trafficking.

August 1990–February 1991
Three squadrons of SAS are deployed to the Gulf as part of British special forces group contribution to the Gulf War. They conduct a range of operations, including reconnaissance and strike actions against Iraqi communications, supply routes and Scud missile launchers.

3 June 1991
Three IRA terrorists are killed in an action by the SAS in Coagh, County Tyrone.

1994–
SAS forces are deployed to the former Yugoslavia with UN forces. They conduct reconnaissance of Serbian positions and observe possible war-crime activity. One SAS soldier is killed during action with Serbian soldiers.

10 September 2000
SAS soldiers assist in the rescue of seven hostages held by a militia group in Sierra Leone.

September 2001–
SAS forces are believed to have been deployed in Afghanistan in response to terrorist attacks in the USA.

SAS Organization

22nd Special Air Service Regiment and the TA units, 21 and 23 SAS, are part of the UK Special Forces Group, which is commanded by the Director of Advanced Forces, a London-based post. The position was formerly titled the Director of Special Forces.

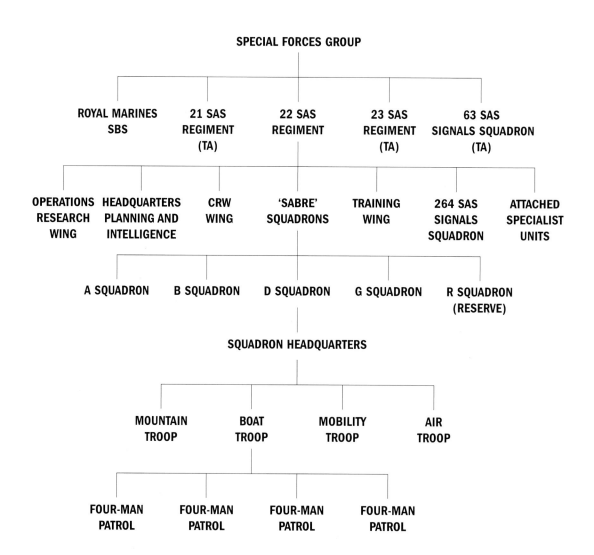

Index

References in italics refer to
illustrations

A

abseiling *95, 176*
action *see* conflict
Aden 186
aggression, controlled 18–21
Air Troops 160, 169–71
AK47 rifles 112
AK74 rifles 113
alcohol consumption 40–41, 88, 137
ambush training 128
amphibious operations 166–9
animals, as food 143–6, 149
anonymity 25
anti-tank weapons 115
application 30–55
aptitude tests 62–3
Arctic Warfare Training Course
    171–2
arms *see* weapons
Auchinleck, General Sir Claude 15

B

backpacks 72–7, 84, *139*
badge *8*
Beckwith, Colonel Charles 7
bergens 72–7, 84, *139*
Billière, General Sir Peter de la *19*,
    27
birds, as food 144
blisters 102
Blood Chits and Blood Money 144
Boat Troops 160, 166–9
boats *see* water craft
boots 71–2
Borneo *148–9*, 186
'Bravo Two Zero' 28, 96–7, 112–13
Browning Hi-Power pistols 8, 111
Browning Machine-guns 17, 163

C

Calvert, Brigadier Mike 28
camouflage 122–5, *137*, 155
camping *67*, 92–3
    see also shelters
carrying equipment 72–7, 84, *139*
characteristics of SAS soldiers
    12–29, 89–91
cheating 91

Chevrolet trucks *16–17*
chin-ups 47
chronology 184–7
Churchill, Winston 14–15
circuit training 51
Clark, Lieutenant-Colonel Dudley
    14–15
Claymore mines 128
clothing 71–2, *73*, 84, *90*
    see also uniforms
Cold War 112
Colditz Castle 18
combat boots 71–2
Combat Survival Training (CST) 107,
    130–57
communications 125–6, *178–9*,
    180–1
    Morse code *122*, 125
compass navigation 63–7, 77, 84
conflict
    Aden 186
    Borneo *148–9*, 186
    Falklands War *132*, *164–5*, 166,
        169, 187
    Gulf War *19*, 25, *26–7*, 27–8,
        28–9, 112–13, 115, 126, *137*,
        144, *158–9*, 162, *162*, *163*,
        164–6, 187
    Iranian Embassy, London *6–7*,
        8–12, 21–2, 25, 187
    Malayan Emergency 22–3, 28, 186
    Northern Ireland *20*, 22, 25, 174,
        187
    Oman 79, 186, 187
    Vietnam War 186, 187
    World War II *13*, 14–21, 184–5
    Continuation Training 61,
        104–29, 130–57
cooking 143
Counter Revolutionary Warfare Wing
    121, 173–7
pistol shooting 160
crunches 45, 47
CST (Combat Survival Training) 107,
    130–57

D

deadfall traps *142*, 144
decision making 52
Democratic Front for the Liberation
    of Arabistan (DFLA) 8–12

demolition experts 181–3
depression 93–6
diet 34–7, 41, *85*, 93
distance runs 50
dogs, evading 155
drinking *see* alcohol consumption
drug abuse 41

E

eating *see* food
ego 23–5
endurance 39, 50–2, *56–7*, 59–61,
    67, 87–91
equipment-carrying 72–7, 84, *139*
Escape and Evasion 107, *114–15*,
    132, 133, 150–55
explosives 181–3
exposure 81, 96–9

F

Falklands War 166, 187
    clothing *132*
    Klepper canoes *164–5*
    Rigid Raiders 169
'Fan dance' 85–6
fighting *see* conflict
FIM-Stinger surface-to-air missiles
    164
fire ploughs 147
fires *134*, 146–7
first aid *24*, 96–100, 102, *103*
    gunshot wounds *172*
fishing 143
fitness 34–55, 61–2, 83–4
flexibility 39, 49–50
flint and steel, for lighting fires
    146–7
FN FAL rifles 112
food 34–7, *85*, 93
    in hostile terrains 133, 138–46
    four-man patrols 116–20, *122*
    frame charges 9
    frame shelters *146*
    frostbite *98*, 99
    fuel tablets *145*

G

GAU-19 multi-barrelled machine-
    guns 164
General Purpose Machine Gun
    (GPMG) 111, 163

Acknowledgements

Sincere thanks go to Adrian Weale for his kind permission to use extracts of first-hand experiences from his insightful and revealing history of the SAS, *The Real SAS* (Pan, London, 1999). *The One That Got Away* by Chris Ryan is published by Arrow Books and extracts are reprinted by permission of The Random House.

Picture Credits

Corbis: 98, 152. Military Picture Library International: 36, 62, 65, 68, 78-79, 104-105, 110, 123, 128, 130-131, 135, 145, 154, 157, 162, 164-165, 168-169, 178-179, 181, 182-183. All other pictures supplied by TRH Pictures / Private Collections